NANO-BIOCHAR AND NANOPARTICLES FOR SOIL HEALTH IMPROVEMENT

NANO-BIOCHAR AND NANOPARTICLES FOR SOIL HEALTH IMPROVEMENT

Vishnu D. Rajput, PhD
Tatiana M. Minkina, PhD
Editors

APPLE ACADEMIC PRESS

First edition published 2025

Apple Academic Press Inc.
1265 Goldenrod Circle, NE,
Palm Bay, FL 32905 USA

760 Laurentian Drive, Unit 19,
Burlington, ON L7N 0A4, CANADA

CRC Press
2385 NW Executive Center Drive,
Suite 320, Boca Raton FL 33431

4 Park Square, Milton Park,
Abingdon, Oxon, OX14 4RN UK

Library and Archives Canada Cataloguing in Publication

CIP data on file with Canada Library and Archives

Library of Congress Cataloging-in-Publication Data

CIP data on file with US Library of Congress

ISBN: 978-1-77491-854-8 (hbk)
ISBN: 978-1-77491-855-5 (pbk)
ISBN: 978-1-00360-386-3 (ebk)

About the Editors

Vishnu D. Rajput, PhD
Leading Researcher (Associate Professor),
Head of International Laboratory of Nanobiotechnology,
Southern Federal University

Vishnu D. Rajput, PhD, is working as a leading researcher (Associate Professor) and Head of the Soil Health Lab at Southern Federal University, Rostov-on-Don, Russia. His ongoing research is based on the toxic effects of bulk- and nano-forms of metals and investigating the bioaccumulation, bio/geo-transformations, uptake, translocation, and toxic effects of bulk- and nano-forms of metals on plant physiology, morphology, anatomy, the ultrastructure of cellular and subcellular organelles, cytomorphometric modifications, and DNA damage. He is also researching possible remediation approaches such as using biochar/nano-biochar-based sorbents and nanotechnological approaches. With long experience and experimental work, Dr. Rajput comprehensively detailed the state of research in environmental science regarding how nanoparticles/heavy metals interact with plants, soil, microbial community, and the larger environment as well as possible remediation technology using nanoparticles/nano-biochar. He is an internationally recognized reviewer, having reviewed 205 manuscripts; he received Outstanding Reviewing Certificates from Elsevier, Springer, and the journal *Plants*. He is an editorial board member of various high-impact journals. He is a leader of several grants including Mega-Grants and BRICS grants and performer/co-principal investigator of 21 national and international projects. Dr. Rajput has received Certificates of Appreciation (2019, 2021) Certificate of Honor (2020), Diploma Awards (2021 and 2022), and Letter of Gratitude (2022) from Southern Federal University, Russia, for his outstanding contribution to academic, creative research, and publication activities. He has also received "Highly Qualified Specialist" status from the Russian government.

Tatiana M. Minkina, PhD
Head, Department of Soil Science and Land Evaluation, Southern Federal University

Tatiana M. Minkina, PhD, is the Head of the Soil Science and Land Evaluation Department of Southern Federal University. She is also the Head of the International Master's Degree Educational Program "Management and Estimation of Land Resources" (2015–2022, accredited by ACQUIN). Her areas of scientific interest include soil science, biogeochemistry of trace elements, and environmental soil chemistry. She focuses on soil monitoring, assessment, modeling, and remediation using physicochemical treatment methods. In 2015, she was awarded a Diploma from the Ministry of Education and Science of the Russian Federation for her long-term work in developing and improving the educational process and her significant contribution to training highly qualified specialists. Currently, she is involved in projects funded by the Russian Scientific Foundation, Ministry of Education and Science of the Russian Federation, and Russian Foundation of Basic Research. She is a member of the Expert Group of the Russian Academy of Science, the International Committee on Contaminated Land, Eurasian Soil Science Societies, the International Committee on Protection of the Environment, and the International Scientific Committee of the International Conferences on Biogeochemistry of Trace Elements. She has a total of 757 scientific publications (389 in English), with 141 in the Web of Science and 195 in Scopus. Her citation index is 803 in the Web of Science with an h-index of 16, and 1281 in Scopus with an h-index of 19. She is an invited editor of the open access journal by MDPI *Water*. Additionally, she serves as an editorial board member of *Geochemistry: Environment, Exploration, Analysis* and *Eurasian Journal of Soil Science.*

ACKNOWLEDGEMENT

VDR acknowledge the support by the Strategic Academic Leadership Program of Southern Federal University, known as "Priority 2030."

Contents

Contributors ... *ix*

Preface ... *xi*

1. **Fabrication and Characteristics of Nano-Biochar: Recent Technological Updates ... 1**
   Gurpreet Singh, Vishnu D. Rajput, and Neetu Sharma

2. **Application of Nano-Biochar to Improve Soil Quality and Sustainability ... 33**
   Shreni Agrawal, Pradeep Kumar, Richa Das, Amit Kumar Singh, Praveen Kumar Shukla, Pooja Verma, Vishnu D. Rajput, Indrani Bhattacharya, Sunil Kumar Mishra, and Kavindra Nath Tiwari

3. **Nanoparticle, Types, and Approaches for Improving Soil Health ... 59**
   Richa Das, Pradeep Kumar, Shreni Agrawal, Amit Kumar Singh, Praveen Kumar Shukla, Pooja Verma, Vishnu D. Rajput, Indrani Bhattacharya, Sunil Kumar Mishra, and Kavindra Nath Tiwari

4. **Nano-Biochar: A Review on Its Applications in Agriculture Soil Productivity ... 95**
   Alok Bharadwaj, Akanksha Shakya, and Mansi Gupta

5. **Role of Nanoparticles and Nano-Biochar in Reducing Salinity Stress ... 117**
   Nihad Ashraf Khan

6. **Application of Nanoparticles and Nano-Biochar for Improving Soil Physiochemical Properties ... 153**
   Sheetal Kumari and Manoj Chandra Garg

7. **Insights into Nanoparticles and Nano-Biochar in Soil-Microbe Interaction ... 177**
   Ritika Chauhan, Gaurav Yadav, Ajit Varma, D. K. Choudhary, and Arti Goel

8. **Role of Nanoparticles and Nano-Biochar in Improving Intrinsic Soil Characteristics** **205**
Fakher Abbas, Mahmood Ul Hassan, Jianjun Du, Aleeza, Rida Bulbul, and Ruqin Fan

Index ***239***

Contributors

Fakher Abbas
Guangdong Engineering and Technology Center for Environmental Pollution Prevention and Control in Agricultural Producing Areas, College of Environmental Science and Engineering, Zhongkai University of Agriculture and Engineering, Guangzhou, China

Shreni Agrawal
Department of Biotechnology, Parul Institute of Applied Science, Parul University, Vadodara, Gujarat, India

Aleeza
Institute of Soil and Environmental Sciences, PMAS-Arid Agriculture University Rawalpindi, Pakistan

Alok Bharadwaj
Department of Biotechnology, GLA University, Mathura, Uttar Pradesh, India

Indrani Bhattacharya
Department of Biotechnology, Parul Institute of Applied Science, Parul University, Vadodara, Gujarat, India

Rida Bulbul
Institute of Horticulture, PMAS-Arid Agriculture University Rawalpindi, Pakistan

Ritika Chauhan
Amity Institute of Microbial Technology, Amity University, Noida, Uttar Pradesh, India

D. K. Choudhary
Amity Institute of Microbial Technology, Amity University, Noida, Uttar Pradesh, India

Richa Das
Department of Biotechnology, Parul Institute of Applied Science, Parul University, Vadodara, Gujarat, India

Jianjun Du
Guangdong Engineering and Technology Center for Environmental Pollution Prevention and Control in Agricultural Producing Areas, College of Environmental Science and Engineering, Zhongkai University of Agriculture and Engineering, Guangzhou, China

Ruqin Fan
Guangdong Engineering and Technology Center for Environmental Pollution Prevention and Control in Agricultural Producing Areas, College of Environmental Science and Engineering, Zhongkai University of Agriculture and Engineering, Guangzhou, China

Manoj Chandra Garg
Amity Institute of Environmental Science (AIES), Amity University, Noida, Uttar Pradesh, India

Arti Goel
Amity Institute of Microbial Technology, Amity University, Noida, Uttar Pradesh, India

Mansi Gupta
Department of Biotechnology, GLA University, Mathura, Uttar Pradesh, India

Mahmood Ul Hassan
College of Grassland Agriculture, Northwest A&F University, Yangling, China

Nihad Ashraf Khan
Department of Biosciences, Jamia Millia Islamia University, New Delhi, India

Pradeep Kumar
Department of Botany, MMV, Banaras Hindu University, Varanasi, Uttar Pradesh, India

Sheetal Kumari
Amity Institute of Environmental Science (AIES), Amity University, Noida, Uttar Pradesh, India

Sunil Kumar Mishra
Department of Pharmaceutical Engineering and Technology, Indian Institute of Technology, Banaras Hindu University, Varanasi, Uttar Pradesh, India

Vishnu D. Rajput
Academy of Biology and Biotechnology, Southern Federal University, Rostov-on-Don, Russia

Akanksha Shakya
Department of Biotechnology, GLA University, Mathura, Uttar Pradesh, India

Neetu Sharma
Biological Systems Engineering, Plaksha University, Mohali, Punjab, India

Praveen Kumar Shukla
Department of Botany, MMV, Banaras Hindu University, Varanasi, Uttar Pradesh, India

Amit Kumar Singh
Department of Pharmaceutical Engineering and Technology, Indian Institute of Technology, Banaras Hindu University, Varanasi, Uttar Pradesh, India

Gurpreet Singh
Biological Systems Engineering, Plaksha University, Mohali, Punjab, India

Kavindra Nath Tiwari
Department of Botany, MMV, Banaras Hindu University, Varanasi, Uttar Pradesh, India

Ajit Varma
Amity Institute of Microbial Technology, Amity University, Noida, Uttar Pradesh, India

Pooja Verma
Department of Botany, MMV, Banaras Hindu University, Varanasi, Uttar Pradesh, India

Gaurav Yadav
Amity Institute of Microbial Technology, Amity University, Noida, Uttar Pradesh, India

Preface

Healthy soils are key to ensuring food security in a sustainable way. However, soil degradation, pollution, and loss of soil productivity are global issues, representing major threats to meeting food requirements. A large number of technological inventions have been applied to clean up soil and restore soil fertility; however, due to being expensive, unfeasible, and labor-consuming in realistic environments or larger applications, past approaches have failed to improve soil health.

Recently, nanotechnical approaches have gained recognition in soil health improvement, especially nano-sized biochar and various metal-based nanoparticles, which may have a direct influence on soil microbial functionality, soil biota, and soil physiochemical properties. Fabrication of nano-sized composites like biochar or nanoparticles enhances the porosity and surface area, and augments the functional groups for specific interaction with soil systems. Manipulating the soil microbial community either by applying nanoparticles or nano-sized biochar has garnered great attention from the scientific community to cope with adverse environmental stress, sustain soil health, and maintain soil fertility. Degraded soils can impose various types of stress on plants. Nanotechnological approaches can help understand the plant-microbe mechanism interaction, reduce soil stresses on plants, and improve soil health. However, some aspects are not well-explored, such as how nanotechnology could improve or manipulate rhizospheric microbiome structure and functionality, as limited studies are available. Therefore, a number of experimental outcomes to explore the exact role of nano-biochar and nanoparticles individually and in combination are needed to improve soil health via enhancing soil physiochemical properties, managing soil stresses such as abrupt rises in heat, coping with drought and salinity stress, improving nutrient availability, reducing toxicity, and sustainably restoring soil fertility.

This new book, *Nano-Biochar and Nanoparticles for Soil Health Improvement*, deals with the crucial issues of soil health and its improvement, especially in a climate-changing environment. Improvement in soil health will help to ensure food security. The book explores how nanotechnology is showing promising results for enhancing soil health and for achieving sustainable agriculture. Various nanotechnological approaches have been applied; however, recently biochar, nano-biochar, as well as nanomaterials have played significant roles in improving soil health.

In this book, the focus is on the fabrication and characteristics of biochar and nano-biochar with recent technological developments, the application of these materials to improve soil quality and sustainability, nanoparticles and their types, approaches for improving soil health, uses to enhance soil productivity, reducing soil salinity, and improving soil physiochemical properties, as well as insights detailing on soil-microbe interaction.

This book is designed to address solutions to soil stress and to improve soil quality and soil biota for sustainable and enhanced crop production. This comprehensive information will be useful for researchers, academicians, policymakers, scholars, students, and farmers to understand the soil issues and their remedies to enhance soil productivity. It will be a valuable addition to the scientific community to explore more insights into technological developments to improve soil health and meet the required food demand.

—Vishnu D. Rajput, PhD
Rostov-on-Don, Russia

CHAPTER 1

Fabrication and Characteristics of Nano-Biochar: Recent Technological Updates

GURPREET SINGH,[1] VISHNU D. RAJPUT,[2] and NEETU SHARMA[1]

[1]*Biological Systems Engineering, Plaksha University, Mohali, Punjab, India*

[2]*Academy of Biology and Biotechnology, Southern Federal University, Rostov-on-Don, Russia*

ABSTRACT

With the rapid transformation of the world, multiple challenges have arisen for the environment in terms of climate change, and toxicity of soil, water, and air. These changes call for revisiting the conventional methods for tackling the issues posed by technological interventions. Carbon is one of the most abundant elements on earth with multifunctional applications in agriculture, industry, medicine, etc. It is one of the major candidates of interest for nanotechnology researchers too. Biochar is one of the products produced by anaerobic thermal treatment using a range of natural and waste materials like straw, corn waste, wood, municipal solid waste, etc. The raw material and treatment temperature are the driving factors for determining the final physicochemical properties of biochar. N-BC, having the properties of biochar with nano intervention, is grabbing the attention of all researchers owing to its potential to provide sustainable solutions to environmental challenges along with climatic change by acting as a sink for excess carbon through sequestration from the atmosphere. N-BC has an

Nano-Biochar and Nanoparticles for Soil Health Improvement.
Vishnu D. Rajput and Tatiana M. Minkina (Eds.)

edge over conventional biochar due to its small size and enhanced properties like specific surface area, microporosity, cation exchange capacity, etc. N-BC is produced from biochar using a conventional milling process or by flash heating. The characteristics and properties of N-BC are primarily dependent on the processing method of raw material. The research is still in its infancy and the associated factors need to be explored. For example, its ability to be easily mobilized calls for risk assessment in aquatic and terrestrial living systems and its potential impact on animals and humans. This chapter is an extensive compilation of the studies on the production, characterization, applications, and recent developments in the field of N-BC research and their prospects for the world.

1.1 INTRODUCTION

The increasing environmental awareness globally calls for the adoption of sustainable approaches in all fields of life. The rise in demand for available energy resources, such as fossil fuels, to meet different needs in industrial and agricultural sectors is not only putting extra pressure on the available stock but also impacting the environment through the addition of greenhouse emissions and by-product accumulation. To ease the current burden, there is an urgent need to develop a circular economy and sustainable options. Carbon-based materials such as "Biochar" (BC) have the potential to provide solutions to these challenges. Biochar, a product formed through the processing of organic waste biomass (animal waste, plant residues, agricultural and municipal solid waste), has multiple applications ranging from soil amendment to acting as remediation agents for the removal of contaminants and heavy metals from soil and water. Furthermore, biochar has been widely used as an animal feed supplement, as a porous material for improving soil health and managing greenhouse gases, nutrient, and microbial carriers, immobilizing agent for the remediation of toxic metals and organic contaminants in soil and water, and as a catalyst for industrial applications [1, 2].

The high adsorptive potential and presence of functional groups facilitate the entry of BC in emerging fields like healthcare industries, microbial fuel cells, and in the manufacture of sensors and supercapacitors [3, 4]. The final range of biochar particle size, i.e., cm to mm is governed by the pyrolysis process [5]. Based on their size, biochar can be classified into Bulk Biochar (B-BC) having a size in the range of 0.04–20 mm, and Nano-Biochar (N-BC) with a size smaller than 100 nm. The former undergoes physical degradation to form N-BC [6, 7]. Further developments have led to the development of N-BC having a diameter less than 5 nm which is an asset to many fields [8]. The favorable structural dimensions and improved physicochemical properties make N-BC a preferred area of research owing to its multiple applications across various fields [9]. Other important characteristics of N-BC are the soluble and insoluble phases and the ability to decrease or increase the soluble content of carbon in these by adjusting the pH and pyrolysis temperature which has further widened its area of application in the agricultural and environmental sciences [7].

N-BC can be manufactured by top-down and bottom-up approaches, which include a list of methods like laser ablation and arc discharge under the former, and chemical precipitation and pyrolysis under the latter. Most of the available methods suffer from the drawback of high-cost inputs in terms of technology involved, raw materials, and energy requirements [10]. Another area of concern is the lack of toxicity data related to the use of N-BC, which is not readily available. However, toxic behavior has been observed in their close counterparts such as carbon nanotubes (CNT) and graphene oxides towards living organisms [11, 12]. Similarly, N-BC may also pose a risk to flora and fauna and the life associated with it due to its popularization and increased application. Hence, there is a need to emphasize the safety aspects of their usage, track their movement in the ecosystem post-application, and develop cost-effective and environmentally friendly approaches for their manufacture.

The increased interest in the area of N-BC research has resulted in research and review publications in this area. The objective of the current chapter is to summarize the important studies provide detailed insights into the fabrication and characteristics of N-BC, along with potential gaps in the area, and examine future research opportunities holistically.

1.2 FABRICATION OF N-BC

The size reduction of biochar particles to the nanoscale can significantly enhance their properties, such as a higher surface-to-volume ratio, which boosts the surface energy and hence makes it biologically effective [13]. The majority of the methods discussed above have certain drawbacks.

Therefore, the production of N-BC must be made cheap, efficient, and environmentally friendly [10]. The process of fabrication begins with the conversion of biomass to bulk biochar, which is then subjected to size reduction using various separation methods in order to produce a wide range of fine-scale particles and/or nanopowders of biochar (Figure 1.1).

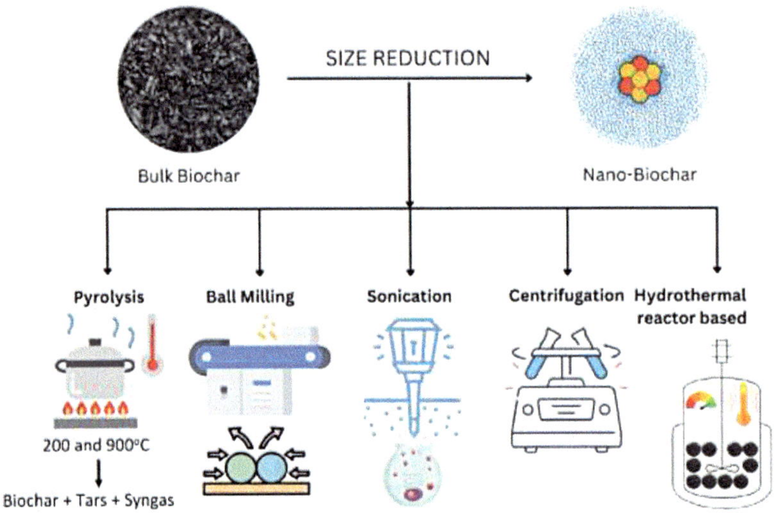

FIGURE 1.1 Various approaches for the fabrication of nano-biochar.

1.2.1 PYROLYSIS

Pyrolysis is a thermochemical process employed for the activated and sustained decomposition of biomass into biochar, tars, and syngas (a mixture of carbon dioxide, carbon monoxide, hydrogen, and light hydrocarbons) under oxygen-free conditions at temperatures between 200 and 900°C [14]. Depending on the biomass feedstock, temperature, heating rate, and residence time, the product yield and carbon distribution may vary in carbonaceous materials. Based on the operating conditions such as heating rate, pyrolysis can be broadly categorized as slow and fast pyrolysis, catalytic pyrolysis, flash pyrolysis, microwave pyrolysis, hydrothermal carbonization, and torrefaction. Slow pyrolysis is the standard thermal conversion method employed to obtain high-yielding biochar. The biomass feedstock is heated slowly from the optimal temperature to the treatment temperature (usually > 400°C) in the presence of an inert gas to prevent the oxidation of carbon. The longer residence time in this process allows for the polymerization reaction of condensable volatile compounds to produce biochar. Fast pyrolysis and flash pyrolysis occur at high heating rates with short vapor residence time compared to those of slow pyrolysis. Flash pyrolysis is regarded as a better process for efficiently synthesizing biochar and syngas [15].

In microwave pyrolysis, irradiated biomass components exhibit movement in terms of ion migration or dipole rotation, leading to the generation of heat because of the friction among the molecules. This in situ heating process significantly restricts the mass transfer of volatile elements in the biomass and stimulates a continuous reaction leading to the formation of biochar. However, the yield of biochar is comparable to that obtained after slow pyrolysis [16].

Hydrothermal carbonization and torrefaction usually take place at lower temperatures (< 350°C), yielding biochar as the major product with gas and liquid as by-products in low amounts. The process of biomass carbonization occurs in autoclave reactors or pressure vessels, which operate efficiently at conditions of high pressure and temperature and are capable of effectively converting biomass feedstock materials without the input of additional energy [17, 18].

1.2.2 GASIFICATION

Gasification is a thermochemical conversion method that involves converting carbonaceous materials into syngas using gasification agents. Syngas is the main product of gasification, with tar, char, and ash as byproducts. Air, water vapor, oxygen, and their mixtures are the most commonly used gasification agents. This process produces hydrogen and carbon monoxide-rich gases by partially oxidizing carbon from pyrolyzed biomass with steam, carbon dioxide, or air at temperatures higher than 700°C [14, 19]. To produce high-energy syngas, biomass must be dried and pyrolyzed first, followed by partial oxidation of carbon. Therefore, gasification generally results in lower yields of biochar than pyrolysis. A partial oxidation condition results in some carbon being converted to CO, which results in low biochar production [20]. Raw biochar can be partially gasified or physically activated to improve its surface chemistry and porosity.

Drying, pyrolysis, oxidation, and reduction are the four steps of gasification. Gasification usually begins with the pretreatment of biomass feedstock. Certain raw materials must be dried in advance because of their high water content [21, 22]. Various factors affecting the characteristics and composition of biochar and gas include the gasifier type, feedstock materials, and operating conditions (pressure, temperature, and residence time) [23]. Temperature plays a key role in determining the efficiency of gasification, biochar and gas composition, and the formation of tar. Codignole et al. [24] studied the impact of different gasification temperatures on the properties of biochar and found that high temperatures produced polycyclic aromatic hydrocarbons (PAHs) and alkali in the non-gasified biochar. The gasification agent has a significant effect on the composition of products as well as their conversion rate. Valizadeh [25] discovered that the biohydrogen produced by steam gasification of food wastes is less reactive and more compressed than biochar produced by air gasification. In general, gasification produces combustible syngas from biomass in addition to biochar. Due to the higher processing temperature, gasified biochar has reduced porosity values than pyrolytic biochar [26]. A few studies have investigated the potential of gasified biochar to capture CO_2 [27, 28]. Dissanayake et al. [29] synthesized biochar by the gasification of chicken manure and wood chips and attained a higher degree of adsorption.

1.2.3 BALL MILLING

Balling has proven to be the most effective method for the production of N-BC. Ball milling has become increasingly popular as a greener, reproducible, and more economical method of manufacturing fine-scale biochar-based materials with enhanced properties and adaptability to a wide range of applications [30]. It is a top-down process for producing N-BC in which the biochar particles are disintegrated into nano-size by collision between metallic balls [31]. An energy mill is used in this process along with a powder charge and a milling medium [32]. When moving balls move across the powder charge, the kinetic energy and mechanical stress generated stretch or break down chemical bonds between molecules and disrupt the lattice structure of the materials, resulting in smaller particles [33]. The particles can be sized up to grinding limits by reducing the rate of agglomeration and changing the ball milling conditions (ball sample ratio and milling time). Ball milling also eliminates the need for hazardous chemicals, which reduces the amount of waste generation, allows for biodegradability, and utilizes renewable resources [30, 34–38]. There are two methods of ball milling biochar to the nanoscale, namely, Wet milling and Dry milling. There are more advantages to wet milling than dry milling owing to the fact that it yields smaller particles with enhanced dispersibility, more functional groups, and a greener technology that saves time and labor [39, 40]. Naghdi et al. [35] studied the production of N-BC from pine wood biochar using a planetary ball mill and investigated the adsorption of carbamazepine on raw and N-BC to determine the impact of size reduction on the adsorption capacity of biochar, which was found to be higher in the case of N-BC. The surface energy and size of the particles are influenced by several parameters, including feedstock materials, grinding media, temperature, type of mill, rotation speed, substrate loading, milling time, the size, number, and weight of balls, and ball to powder mass ratios [38, 41, 42]. In a study by Ng et al. [43], it was found that mixing steel balls of different sizes (60NBC) was more effective than using small steel balls or large steel balls (60NBC-S and 60NBC-L) of uniform size for grinding biochar. The particle size distribution of N-BC varied with the milling duration. It was reported that biochar that had been mechanically ground and unmilled was approximately 331 nm in size. Ball milling for 30 minutes reduced the diameter of biochar to 90–100 nanometers. N-BC decreased in size as the ball milling duration increased up to 180 minutes. According to Amusat et al. [44], ball milling can produce functional nano-composites by reducing size and modifying the surface properties.

Although planetary ball milling can effectively disintegrate and reduce biochar to nanoscale, the process is time-consuming and requires several hours of pre-conditioning. Another effective ball milling technique known as high-energy milling involves moving the vials and discs in a figure-of-eight pattern, which combines horizontal and vertical movements of the grinding media. This is in contrast to planetary ball milling, in which the grinding media moves counter-rotationally parallel to the ground [45]. Compared to planetary ball milling, which is more commonly used to produce N-BC, high-energy milling generates a considerable impact force on the sample material due to its wide range of motion and high frequency [46]. A high-energy ball mill, however, has been used in limited studies to produce N-BC. In an alternative to conventional planetary ball milling, Ng et al. [43] used high-energy ball milling to produce NBC from oil palm biomass in a shorter period of time with no pre-treatment or pre-conditioning.

Materials created using this method have a higher defect density and crystal deformation when treated at a higher temperature. As a result of the generated energies, bonds may be fractured and radicals may be produced in a variety of ways [47, 48]. Among the key challenges to achieving optimal ball milling performance is the formation of dense agglomerates of nanoparticles. Prior to ball milling biochar, Naghdi et al. [35] suggest pre-conditioning it under –80°C to minimize particle agglomeration. In the same milling conditions, the pre-treatment resulted in reduced agglomeration of particles and a size reduction from 212 to 60 nm. It is also possible to suppress agglomerates with wet milling, or by using solvents such as water and ethanol [49, 50]. Despite its ability to reduce size agglomeration, pre-conditioning biochar increases the N-BC production costs and thus prevents it from being scaled up. The distribution of particles can also be improved by adding magnetic materials such as iron oxides to biochar [36, 51].

1.2.4 SONICATION

Biochar can also be physically activated with adsorption capacity by using ultrasound irradiation. As a result of sonication, biochar in the suspension disintegrates under the influence of high-energy ultrasound [52]. During the process of sonication, shockwaves open the blocked pores in biochar and exfoliate the carbon structure, thereby increasing the microporous area. Consequently, mineral matter can be more readily leached into the pores, resulting in better adsorption [53]. Sonication leads to the formation of N-BC by exfoliating the fine fragments anchored to bulk biochar surfaces or embedded in the pores [7]. This method has the major advantage of forming pores without clogging the surface of biochar thus resulting in a uniform surface [54, 55]. During sonication, the properties of N-BC are altered by changes in temperature, power, probe size, and amplitude. This process produces nanoparticles that are homogeneous in shape and highly pure and have the advantage of being controlled by the reaction conditions and shorter reaction times. This process suffers from certain disadvantages, such as environmental and health hazards, non-uniform energy intensity distribution, and increased manufacturing costs, which may limit its scale-up [56].

1.2.5 CENTRIFUGATION

A simple and effective method for separating and isolating N-BC is sedimentation. This method relies on the sedimentation properties of solid particles under centrifugal or normal conditions [57, 58]. The separation of nanoparticles requires multiple centrifugation rounds. To achieve effective sedimentation of larger particles under static conditions, sedimentation may take up to 24 hours [59]. By centrifuging the biochar solution after dispersing it into a solvent, the sedimentation time can be shortened. As a result of centrifugation, nanoparticles remain in the supernatant, while large particles settle to the bottom of the pellet. For example, Lian et al. [60] and Yue et al. [61] extracted rice husk and rice straw-derived N-BC with sizes ranging from 50 to 280 nm and 100 to 350 nm, respectively, by centrifugation at 10,000 rpm for 30 min. Xiao et al. [62] obtained N-BC with a size range of 100 and 2,500 nm by low-speed centrifugation at 9,000 rpm using milled cow bone meal biochar. In a recent study by Xu et al. [42], N-BC was fractionated from a variety of feedstock materials by centrifugation at 9,500 rpm for 20 minutes. In another study, Placido et al. [63] centrifuged size-reduced bulk biochars of dairy manure, rice straw, and sorghum straw for 20 minutes to isolate fractions of N-BC. By centrifuging hand-milled bagasse biochar at 500 rpm for 3 minutes, Dong et al. [64] obtained N-BC with sizes ranging from 50 to 570 nm. It has been found that the process of centrifugation produces highly stable colloids of N-BC; however, the yield of these colloids is low, and hazardous solvents are used in this process [56].

1.2.6 HYDROTHERMAL REACTOR-BASED

A facile procedure for synthesizing biochar nanodots using hydrothermal reactors is also employed for the fabrication of N-BC. When bulk biochar is exposed to oxidizing, high-pressure and temperature conditions, or aqueous environments, its microstructure is disrupted by acids such as HNO_3 and H_2SO_4, resulting in the formation of N-BC. Hydrothermally produced nanodots are smaller in size than those produced by other methods, such as grinding, which is energy-intensive, and sonication, which produces particles with larger sizes. This method serves as a tool to determine the optimal conditions required for the synthesis of N-BC by systematically adjusting variables such as time, temperature, and aqueous conditions during hydrolysis. Hence, this method can provide simple, quick, sustainable, and cost-effective alternatives to produce biochar nanodots, which can be further used as multifunctional adsorbents and for studying their environmental fate in soil and water [65].

1.3 CHARACTERISTICS

The varied characteristics of BCs make them a preferred choice for various applications. The final properties of the product, such as particle size, specific surface area (SSA), cation exchange capacity, redox potential, moisture, ash and carbon content, porosity, elemental composition, electrical conductivity, functional groups, etc., are primarily driven by the feedstock raw material and the procedure employed [35, 66, 67]. The production strategy for N-BC is very crucial and plays an important role in determining the final composition and application of the final product (Table 1.1). The functional moieties of the end-product are directly dependent on the transformations undergone by the raw materials like pyrolysis temperature, grinding speed and time, chemical augmentation, etc. [7].

The physicochemical characterization of N-BC is carried out by employing different techniques depending on the target to be analyzed. An elemental analyzer was employed to detect the presence of carbon, hydrogen, oxygen, and nitrogen. The morphological characterization was done using transmission electron microscopy, scanning electron microscopy, and atomic force microscopy. Further characterization was done employing X-ray diffraction, Fourier transform infrared spectroscopy, and X-ray photoelectron spectroscopy [7, 90].

1.4 PHYSICAL CHARACTERISTICS

As discussed, the porous structural morphology of N-BC is primarily dependent on the architecture of the raw material. The porosity, in turn, governs the adsorption and water-holding characteristics of biochar. Scanning electron microscopy provides detailed information about the structural details, which further helps in the evaluation of its application area. The temperature range of pyrolysis plays an important role in yielding a honeycomb-like structure in BC, resulting in a high Brunauer, Emmett, and Teller (BET) surface area [91]. The BET analysis further helps in understanding the available SSA, pore size, and volume. The presence of micropores with a diameter of less than 2 nm results in a higher surface area and increased adsorption of N-BC [92]. These properties aid in the increased sorption of materials like pesticides, humic acids, etc. [93].

TABLE 1.1 Sources, Methods, and Applications of Nano-biochar and Biochar Nanocomposites

Type		Method	Feedstock Materials	Temperature (°C)	Applications	References
Magnetic BC	Magnetic biochar (hematite modified)	Pyrolysis	Hematite or goethite mineral and pine needle biomass.	600	Removal of contaminants like sulfamethoxazole.	[68]
	Magnetic corn stover biochar	Pyrolysis	Corn stover	500	Fluoride removal from contaminated groundwater.	[69]
	Calcium-based magnetic biochar.	Pyrolysis	Rice straw	400	Remediation of soils co-contaminated with arsenic and cadmium.	[70]
	Ball-milled magnetic N-BC	Ball milling	Wheat straw	700	Removal of mercury and tetracycline.	[36]
Metal oxide N-BC	Copper oxide modified nanocomposites.	Ball milling	Hickory wood chips	600	Removal of reactive red 120.	[71]
	Iron oxide permeated mesoporous N-BC.	Chemical amended pyrolytic approach.	Rice husk	600	Removal of arsenic	[51]
	Manganese oxide modified biochar, birnessite.	Pyrolysis	Pinewood	600	Removal of arsenic and lead (Pb).	[72]
–	ZnO/ZnS modified biochar	Pyrolysis	Corn stover	500–800	Removal of Pb (II), copper (Cu (II)), and chromium (Cr (VI)).	[73]
	Calcium oxide	Pyrolysis	Ramie stem	500	Removal of phosphate	[74]

TABLE 1.1 *(Continued)*

Type		Method	Feedstock Materials	Temperature (°C)	Applications	References
Clay BC	Clay-biochar composite	Pyrolysis	Local clay and cottonwood.	650	Removal of glyphosate from aqueous solutions.	[75]
		Pyrolysis	Bamboo, bagasse, and hickory chips.	600	Removal of methylene blue.	[76]
		Pyrolysis	Reed straw and clay	600	Removal of NH_{4+} – N from aqueous solution.	[77]
Wood BC	–	Pyrolysis	Pitch pine	600	Applications like bio-fertilizers, briquette biofuels, and Pb_{2+} bio-sorbents.	[78]
		Disc milling	Dendro	700–1,000	Removal of oxytetracycline, glyphosphate, Cr(VI), and Cd(II) from aqueous environment.	[50]
		Pyrolysis	Pecan, cherry, and applewood	350, 500, 700, 900°C	Removal of glyphosphate from aqueous environment.	[79]
		Microwave pyrolysis	Spruce softwood chips and hemp stalk biomasses.	2,100, 2,400, and 2,700 W	Composite filler material.	[80]
Plant waste-derived BC	–	Pyrolysis	Modified cassava waste biochar.	500	Removal of norfloxacin, sulfamerazine, and oxytetracycline.	[81]
		Pyrolysis	Peanut shells and rice husks.	500	Characterization of physicochemical properties.	[59]
		Microwave pyrolysis	Palm kernel shell	600	Removal of heavy metals from wastewater.	[82]
		Pyrolysis	Wheat straw	700	Characterization and comparison of macro- and nano-biochar.	[83]
		Pyrolysis	Rice hull	300–600	The effect on rice plant growth and the uptake of heavy metals.	[61]

TABLE 1.1 *(Continued)*

Type		Method	Feedstock Materials	Temperature (°C)	Applications	References
Municipal source BC.	–	–	Sewage sludge rice straw	700	Reduction in total/leachable content of heavy metals like Cu, Ni, and Zn.	[84]
			Municipal sewage sludge and tea waste.	300	As an adsorbent for the removal of metal ions from water.	[85]
			Municipal sewage sludge and hazelnut shell.	700	Removal of Cu (II) via adsorption.	[86]
Animal waste derived BC.	Dairy manure	Depolymerization/ chemical oxidation	Dairy manure, rice straw, and sorghum straw.	700	Heavy metal detection in aquatic bodies.	[87]
	Pig manure	Pyrolysis	Pig manure	500	Physicochemical properties characterization.	[59]
	Yak manure	Pyrolysis	Yak manure	350	Removal of heavy metals and methylene blue from aqueous solutions.	[88]
	N-BC	Ball milling	Cow bone meal	300	Removal of Cd (II), Cu (II), Pb (II).	[62]
	Swine manure-rice straw	Co-pyrolysis	Swine manure and rice straw.	400	Soil remediation by heavy metal immobilization.	[89]

The favorable pore size of N-BC is very crucial in developing an association with microbes during anaerobic treatment and aids in the formation of biofilms, which provide a protective barrier to microbes from an acidic environment [24, 94]. Studies have reported that treatments at lower temperatures (300–400°C) lead to the production of N-BCs with a low surface area (5.6–47.2 m_2/g), while higher temperature treatments (450–600°C) resulted in a higher surface area in the range of 342–430 m_2/g [95, 96]. The change in surface area is attributed to the formation of pores at high temperatures as a result of biomass devolatilization accompanied by aliphatic alkyl and ester groups [97]. The removal of pore-blocking substances at high pyrolysis temperatures is one of the contributing factors to increased surface area in N-BC [98].

The surface area of N-BC also depends on the methods employed; in ball milling, the range was from 3.67 to 1,736 m_2/g, centrifugation resulted in 21.7 to 253 m_2/g, while in sonication, this was 0.76 to 264 m_2/g [84]. The varied results in different techniques were due to the reduction in grain size and increase in internal surface area in ball milling [99]. Another study reported an alteration in the surface chemistry of MgO-modified biochar [100].

The increased SSA of N-BC improved the removal of CO_2, H_2S, etc., from the biogas fermenter [101]. Similar studies have reported the sequestration of carbon dioxide using the municipal sludge-derived BC where surface-embedded group I-II A metals also played a crucial role [102, 103]. The temperature of the treatment is very significant for the application of the final product. The raw material processed at low pyrolytic temperature resulted in improved properties for the release of nutrients into the soil by the end product but resulted in hydrophobicity and reduced the ability of water retention [104]. The hardness and ductility of N-BC are indicators of its potential use as a composite filler. N-BC prepared from softwood and hemp were observed to have increased hardness and ductility as observed by Young's modulus. The other contributing factors included a decreased percentage of volatile components and an increased carbon content achieved on exposure to higher temperatures [81].

1.5 CHEMICAL CHARACTERISTICS

The chemical characteristics of N-BC showed variable results depending on the processing temperature and composition of the feedstock. The fixed carbon is the primary combustible residue after the processing of the sample and expulsion of volatile matter, hence can be used to get an estimate of the carbonaceous content of the product [105, 106]. Van Krevelen diagrams were used to estimate the degree of aromaticity and maturation based on H/C and O/C ratios and suggested the effect of the process parameters on the production and chemical composition of the end products [107–109]. The surface area was found to be increased as confirmed by increased BET (Brunauer, Emmett, and Teller) areas due to an increase in carbon burn-off [108]. The ash and carbon content are important components for N-BC application in different fields. The ash content also represents the final pH of the product which in turn depends on its mineral content. The presence of base cations and minerals such as calcium carbonate, silicon dioxide, and elements such as Mg, Al, Na, etc., and volatilization of acidic functional groups resulted in neutral to alkaline pH [50]. Due to its usage as a soil conditioner, the pH of N-BC is very crucial, which is mainly driven by carbonates, the presence of organic compounds in soluble form, and organic functional groups along with some inorganic alkalis [110]. The increase in pyrolysis temperature resulted in increased ash content in N-BC as observed in a goethite-modified peanut shell, which exhibited pH in the range of 6.4–6.8 when treated at 300°C and 400°C, while at 500°C and 600°C, pH rose to 7.08–7.28 [7, 14]. Like other factors, the ash and carbon content of N-BC also depend on the composition of the raw material. For example, the coconut shell-derived N-BC had a higher carbon content compared to the sewage sludge-derived product.

Ball milling resulted in an increased collision between metallic balls and B-BC, which resulted in increased temperature and hence caused volatilization of amorphous carbon, resulting in low carbon and high ash content in N-BC [38]. However, in sonication, the ash content is low due to the separation of minerals and increased oxygen content because of the formation of carboxylic, phenolic, and ketone groups [7]. However, there is no direct correlation between SSA and carbon/ash content; hence, more studies are required to draw a correlation between total carbon and ash content with the surface area of N-BC [56].

The FTIR spectra indicate that biochar is dominated by functional groups typical of oxygenated hydrocarbons, reflecting the carbohydrate structure of cellulose and hemicelluloses [92]. As discussed earlier, increased pyrolysis temperature resulted in the devolatilization of volatile organic compounds, thus resulting in their loss. The functional groups hold an important place in the final product due to their role in nutrient retention and removal of pollutants [32]. Studies reported that ball-milled processed raw material resulted in more oxygen-containing functional groups in the resultant BC as compared to pristine BC, accompanied by an increased number of acidic functional groups, thus resulting in increased adsorption [99]. The type of solvents employed during the ball milling process also played an eminent role in determining the final ratio of functional groups in BC, with water and ethanol resulting in an increased number of acidic groups. N-BC exhibits a variety of functional groups like N–H, SiO_2, –COOC, –COOH, C=C/C=O, and other newly evolved aliphatic groups [80, 99, 111, 112]. The –COOH and OH groups of BC were reported to be responsible for the removal of hydrogen sulfide from wastewater [113]. Cation exchange capacity (CEC) is another important component in deciding the N-BC's application potential in agriculture, as high CEC contributes to improved nutrient exchange and absorption [114]. The reduction of particle size to produce N-BC further expedites the ion exchange [115].

The electron-donating and accepting property of N-BC is ascribed to the presence of free radical groups, and surface functional groups like phenolics, quinoids, metals, and their oxides. The presence of the above moieties and their further controlled oxidation may enhance the redox potential of N-BC and hence its applications [116–118]. The zeta potential is representative of the charge carried by the surface of N-BC, and its higher value is an indicator of a lower level of agglomeration, stability, and higher dispersion [59, 119]. The zeta potential of N-BC reported an increased percentage with respect to B-BC and its parent material [120]. Furthermore, the feedstock also influenced the value of zeta potential with plant-born N-BC having higher values as compared to municipal sources [59]. As a result, lower zeta potential and a lesser number of oxygen-containing groups contribute to a higher number of suspended particles, hence lesser settleability [50].

1.6 APPLICATIONS OF N-BC

Biochar has traditionally been used for bioremediation and soil amendment. N-BC has demonstrated promising potential in the remediation of contaminated soils and water, providing renewable energy alternatives and more recently as an agent to mitigate climate change (Figure 1.2). A variety of processes have been used to create biohydrogen using N-BC, including methane steam reforming, water splitting, and anaerobic digestion [121]. Due to the generation of high-energy active sites in carbonaceous materials doped with metals in trace amounts, more methane can be converted to hydrogen by enhancing methane's attraction to these active sites. Currently, N-BC developed from biosolids derived from wastewater treatment plants is used to reform methane at a 65% conversion rate [122]. Anaerobic digestion of organic bio-waste produces bioenergy. N-BC addition during AD increases hydrogen yield with a minimal lag phase. It is expected that N-BC promotes the formation of biofilms, stabilization of pH, and production of volatile fatty acids [123]. In addition, the minerals present in N-BC act as growth adjuncts for metabolism and enzyme synthesis in microbes. During the process of biofilm formation, N-BC promotes the secretion of extracellular polymeric substances by microbes resulting in increased adherence on the N-BC surface [124].

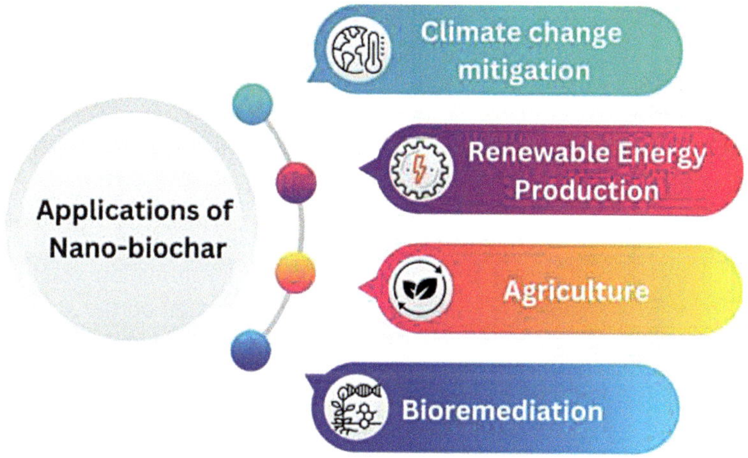

FIGURE 1.2 Environmental applications of nano-biochar.

1.7 CHALLENGES AND RISKS

The wider application of N-BC has made it a favorite candidate in research and development. The nano size and higher dispersion potential bring a new heap of challenges for the researchers. The detailed assessment of the life cycle of N-BC is crucial to prove and attain its long-term sustainable use. Similar to their engineered carbonaceous counterparts, they can pose a risk to flora and fauna [125, 126]. The high adsorption potential also poses another challenge as N-BC can equally sorb both nutrients and contaminants, hence leading to their non-specific dissemination in the environment.

The toxicity of N-BC is governed by its concentration and degree of reactivity in plant and animal systems. Chen et al. [127] observed that the morphology and hydrophobic nature of biochar facilitates its interaction with plant and mammalian cells, leading to their damage. In another study, N-BC showed higher toxicity towards Streptomyces compared to graphene and carbon nanotubes [128]. The higher pyrolysis temperature employed for the production of N-BC resulted in the production of polyaromatic hydrocarbons which, along with trace metals, may pose a risk to exposed mammalian cells [129]. However, there are no specific studies available to establish the toxicity of N-BC to human health. Nevertheless, a life cycle assessment of these particles employing ReCiPe and Tool for Reduction and Assessment of Chemicals is required to assess their production optimization and environmental performance and fate, hence helping in determining their outcome in the focused area of applications [121, 130, 132–149].

1.8 FUTURE PERSPECTIVES

With the increasing interest in eco-friendly techniques, the last decade has witnessed a tremendous boost in the research and application of N-BC across various sectors. The properties of N-BC are primarily driven by the composition of feedstock and processing techniques. Hence, the judicious selection of the two is very critical in determining the quality and application of BC. Most of the studies are concentrated on the application of the final product due to its optimal size and related properties, but there is a need to focus on the optimization of green production processes like ball milling carried at low temperatures along with the scope of scaling up to obtain desired deliverables in required quantity [35]. Similarly, new methods like flash pyrolysis, sonication, and direct production techniques (exfoliated N-BC from corncob) should also be explored to yield a better product. The idea of circular economy can also be inculcated by utilizing organic-rich waste biomass based on scientific knowledge to produce N-BC. More research is required for the characterization of N-BC, and there is a need to establish a relationship among various properties of N-BC like pH, CEC, SSA, SFG, zeta potential, etc., in order to establish a sound base for its future use as current findings are mainly based on the characteristics of bulk biochar. Furthermore, the emerging areas of application like catalysts and high capacitance material in electronics, semiconductors, biosensing, engine additives, etc., call for digging into the required characterization studies [66, 131].

1.9 CONCLUSION

The positive outcomes of N-BC on plant growth and contaminant removal are already established, but its fate and potential toxicity outcome in the environment is a gray area. The transformation of functional groups, metal profiles, and other minerals should be considered of utmost importance as these results will further pave the way for understanding the associated fate and mechanism of N-BC's interaction with plant, microbial, and soil systems. Furthermore, detailed risk assessment studies are required which must include the impact on mammalian cells in order to establish its safer application. Hence, more sustainable choices are necessary if we are to continue to sustain our Earth's ecosystems.

ACKNOWLEDGEMENT

VDR acknowledge the support by the Strategic Academic Leadership Program of Southern Federal University, known as "Priority 2030".

KEYWORDS

- **adsorption**
- **ball milling**
- **biochar**
- **biomass**
- **Brunauer, Emmett, and Teller**
- **carbon nanotubes**
- **cation exchange capacity**
- **feedstock**
- **nano-biochar**
- **nanocomposites**
- **polycyclic aromatic hydrocarbons**
- **specific surface area**
- **sustainable technologies**

REFERENCES

1. Liu, W. J., Jiang, H., & Yu, H. Q. (2015). Development of biochar-based functional materials: Toward a sustainable platform carbon material. *Chemical Reviews, 115*(22), 12251–12285. https://doi.org/10.1021/acs.chemrev.5b00195.
2. Bolan, N., et al. (2021). Multifunctional applications of biochar beyond carbon storage. *Journal of Environmental Quality, 67*(2), 150–200. https://doi.org/10.1080/09506608.2021.1922047.
3. Huggins, T., Wang, H., Kearns, J., Jenkins, P., & Ren, Z. J. (2014). Biochar is a sustainable electrode material for electricity production in microbial fuel cells. *Bioresource Technology, 157,* 114–119. https://doi.org/10.1016/j.biortech.2014.01.058.
4. Genovese, M., & Lian, K. (2017). Polyoxometalate modified pinecone biochar carbon for supercapacitor electrodes. *Journal of Materials Chemistry A, 5*(8), 3939–3947. https://doi.org/10.1039/c6ta10382k.
5. Biochar for Environmental Management. (2015). https://doi.org/10.4324/9780203762264.
6. Spokas, K. A., et al. (2014). Physical disintegration of biochar: An overlooked process. *Environmental Science & Technology Letters, 1*(8), 326–332. https://doi.org/10.1021/ez500199t.
7. Liu, G., et al. (2018). Formation and physicochemical characteristics of nano biochar: Insight into chemical and colloidal stability. *Environmental Science & Technology, 52*(18), 10369–10379. https://doi.org/10.1021/acs.est.8b01481.
8. Xiao, X., & Chen, B. (2017). A direct observation of the fine aromatic clusters and molecular structures of biochars. *Environmental Science & Technology, 51*(10), 5473–5482. https://doi.org/10.1021/acs.est.6b06300.
9. Rajput, V. D., et al. (2022). Nano-biochar: A novel solution for sustainable agriculture and environmental remediation. *Environmental Research, 210,* 112891. https://doi.org/10.1016/j.envres.2022.112891.
10. Xiao, D., Yuan, D., He, H., & Gao, M. (2013). Microwave-assisted one-step green synthesis of fluorescent carbon nanoparticles from ionic liquids and their application as novel fluorescence probe for quercetin determination. *Journal of Luminescence, 140,* 120–125. https://doi.org/10.1016/j.jlumin.2013.02.032.
11. Begum, P., Ikhtiari, R., & Fugetsu, B. (2014). The potential impact of multi-walled carbon nanotubes exposure to the seedling stage of selected plant species. *Nanomaterials, 4*(2), 203–221. https://doi.org/10.3390/nano4020203.
12. Combarros, R. G., Collado, S., & Díaz, M. (2016). Toxicity of graphene oxide on growth and metabolism of Pseudomonas putida. *Journal of Hazardous Materials, 310,* 246–252. https://doi.org/10.1016/j.jhazmat.2016.02.038.
13. Sulaiman, G. M., Mohammed, W. H., Marzoog, T. R., Al-Amiery, A. A. A., Kadhum, A. A. H., & Mohamad, A. B. (2013). Green synthesis, antimicrobial and cytotoxic effects of silver nanoparticles using Eucalyptus chapmaniana leaves extract. *Asian Pacific Journal of Tropical Biomedicine, 3*(1), 58–63. https://doi.org/10.1016/S2221-1691(13)60024-6.

14. Ahmad, M., et al. (2014). Biochar as a sorbent for contaminant management in soil and water: A review. *Chemosphere, 99,* 19–33. https://doi.org/10.1016/j.chemosphere.2013.10.071.
15. Bahng, M. K., Mukarakate, C., Robichaud, D. J., & Nimlos, M. R. (2009). Current technologies for analysis of biomass thermochemical processing: A review. *Analytica Chimica Acta, 651*(2), 117–138. https://doi.org/10.1016/j.aca.2009.08.016.
16. Tengku Yasim-Anuar, T. A., et al. (2022). Emerging application of biochar as a renewable and superior filler in polymer composites. *RSC Advances, 12*(22), 13938–13949. https://doi.org/10.1039/d2ra01897g.
17. Weber, K., & Quicker, P. (2018). Properties of biochar. *Fuel, 217,* 240–261. https://doi.org/10.1016/j.fuel.2017.12.054.
18. Jain, A., Balasubramanian, R., & Srinivasan, M. P. (2016). Hydrothermal conversion of biomass waste to activated carbon with high porosity: A review. *Chemical Engineering Journal, 283,* 789–805. https://doi.org/10.1016/j.cej.2015.08.014.
19. Mohan, D., Sarswat, A., Ok, Y. S., & Pittman, C. U. (2014). Organic and inorganic contaminants removal from water with biochar, a renewable, low cost and sustainable adsorbent – A critical review. *Bioresource Technology, 160,* 191–202. https://doi.org/10.1016/j.biortech.2014.01.120.
20. Cheng, B. H., Zeng, R. J., & Jiang, H. (2017). Recent developments of post-modification of biochar for electrochemical energy storage. *Bioresource Technology, 246,* 224–233. https://doi.org/10.1016/j.biortech.2017.07.060.
21. Yao, Z., You, S., Ge, T., & Wang, C. H. (2018). Biomass gasification for syngas and biochar co-production: Energy application and economic evaluation. *Applied Energy, 209,* 43–55. https://doi.org/10.1016/j.apenergy.2017.10.077.
22. You, S., Ok, Y. S., Tsang, D. C. W., Kwon, E. E., & Wang, C. H. (2018). Towards practical application of gasification: A critical review from syngas and biochar perspectives. *Critical Reviews in Environmental Science and Technology, 48*(22–24), 1165–1213. https://doi.org/10.1080/10643389.2018.1518860.
23. Cha, J. S., et al. (2016). Production and utilization of biochar: A review. *Journal of Industrial and Engineering Chemistry, 40,* 1–15. https://doi.org/10.1016/j.jiec.2016.06.002.
24. Codignole Luz, F., Cordiner, S., Manni, A., Mulone, V., & Rocco, V. (2018). Biochar characteristics and early applications in anaerobic digestion–A review. *Journal of Environmental Chemical Engineering, 6*(2), 2892–2909. https://doi.org/10.1016/j.jece.2018.04.015.
25. Valizadeh, S., et al. (2021). Biohydrogen production from catalytic conversion of food waste via steam and air gasification using eggshell- and homo-type Ni/Al_2O_3 catalysts. *Bioresource Technology, 320,* 124313. https://doi.org/10.1016/j.biortech.2020.124313.
26. Qian, K., et al. (2013). Effects of biomass feedstocks and gasification conditions on the physiochemical properties of char. *Energies, 6*(8), 3972–3986. https://doi.org/10.3390/en6083972.
27. You, S., et al. (2017). A critical review on sustainable biochar system through gasification: Energy and environmental applications. *Bioresource Technology, 246,* 242–253. https://doi.org/10.1016/j.biortech.2017.06.177.

28. Madzaki, H., Karimghani, W. A. W. A. B., Nurzalikharebitanim, & Azilbaharialias. (2016). Carbon dioxide adsorption on sawdust biochar. *Procedia Engineering, 148,* 718–725. https://doi.org/10.1016/j.proeng.2016.06.591.
29. Dissanayake, P. D., et al. (2020). Sustainable gasification biochar as a high-efficiency adsorbent for CO2 capture: A facile method to designer biochar fabrication. *Renewable and Sustainable Energy Reviews, 124,* 109785. https://doi.org/10.1016/j.rser.2020.109785.
30. Kumar, M., et al. (2020). Critical review on biochar-supported catalysts for pollutant degradation and sustainable biorefinery. *Advanced Sustainable Systems, 4*(10), 1900149. https://doi.org/10.1002/adsu.201900149.
31. Deguchi, S., Mukai, S. A., Tsudome, M., & Horikoshi, K. (2006). Facile generation of fullerene nanoparticles by hand-grinding. *Advanced Materials, 18*(6), 729–732. https://doi.org/10.1002/adma.200502487.
32. Lin, X., et al. (2017). Mechanochemistry: A green, activation-free, and top-down strategy for high-surface-area carbon materials. *ACS Sustainable Chemistry & Engineering, 5*(10), 8535–8540. https://doi.org/10.1021/acssuschemeng.7b02462.
33. Boldyreva, E. (2013). Mechanochemistry of inorganic and organic systems: What is similar, what is different? *Chemical Society Reviews, 42*(18), 7719–7738. https://doi.org/10.1039/C3CS60052A.
34. Naghdi, M., et al. (2015). Green and energy-efficient methods for the production of metallic nanoparticles. *Beilstein Journal of Nanotechnology, 6*(1), 2354–2366. https://doi.org/10.3762/bjnano.6.243.
35. Naghdi, M., et al. (2017). A green method for production of nano biochar by ball milling- optimization and characterization. *Journal of Cleaner Production, 164,* 1394–1405. https://doi.org/10.1016/j.jclepro.2017.07.084.
36. Li, R., et al. (2020). Removing tetracycline and Hg (II) with ball-milled magnetic nano biochar and its potential on polluted irrigation water reclamation. *Journal of Hazardous Materials, 384,* 121095. https://doi.org/10.1016/j.jhazmat.2019.121095.
37. Richard, S., Rajadurai, J. S., & Manikandan, V. (2016). Influence of particle size and particle loading on mechanical and dielectric properties of biochar particulate-reinforced polymer nanocomposites. *Journal of Composite Materials, 21*(6), 462–477. https://doi.org/10.1080/1023666X.2016.1168602.
38. Wang, T., Rodriguez-Uribe, A., Misra, M., & Mohanty, A. K. (2018). Sustainable carbonaceous bio filler from Miscanthus: Size reduction, characterization, and potential bio-composites applications. *BioResources, 13*(2), 3720–3739. https://doi.org/10.15376/biores.13.2.3720-3739.
39. Zhu, D., Pan, J., Lu, L., & Holmes, R. J. (2015). Iron ore palletization. In R. Li (Ed.), *Iron Ore: Mineralogy, Processing and Environmental Sustainability* (pp. 435–473). https://doi.org/10.1016/B978-1-78242-156-6.00015-0.
40. Yuan, Y., Zhang, N., & Hu, X. (2019). Effects of wet and dry ball milling on the physicochemical properties of sawdust derived-biochar. *Energy Sources, Part A: Recovery, Utilization, and Environmental Effects, 48*(3), 287–300. https://doi.org/10.1080/10739149.2019.1708751.
41. Shah, U. V., et al. (2015). Effect of milling temperatures on surface area, surface energy, and cohesion of pharmaceutical powders. *International Journal of Pharmaceutics, 495*(1), 234–240. https://doi.org/10.1016/j.ijpharm.2015.08.061.

42. Xu, C. Y., Li, Q. R., Geng, Z. C., Hu, F. N., & Zhao, S. W. (2020). Surface properties and suspension stability of low-temperature pyrolyzed biochar nanoparticles: Effects of solution chemistry and feedstock sources. *Chemosphere, 259, Article* 127510. https://doi.org/10.1016/j.chemosphere.2020.127510.
43. Ng, L. Y. F., Ariffin, H., Yasim-Anuar, T. A. T., Farid, M. A. A., & Hassan, M. A. (2022). High-energy ball milling for high productivity of nano biochar from oil palm biomass. *Nanomaterials, 12*(18), Article 3251. https://doi.org/10.3390/nano12183251.
44. Amusat, S. O., Kebede, T. G., Dube, S., & Nindi, M. M. (2021). Ball-milling synthesis of biochar and biochar–based nanocomposites and prospects for removal of emerging contaminants: A review. *Journal of Water Process Engineering, 41,* Article 101993. https://doi.org/10.1016/j.jwpe.2021.101993.
45. Zolriasatein, A., Shokuhfar, A., Safari, F., & Abdi, N. (2018). Comparative study of SPEX and planetary milling methods for the fabrication of complex metallic alloy nanoparticles. *Micro & Nano Letters, 13*(4), 448–451. https://doi.org/10.1049/mnl.2017.0608.
46. Aguilar, C., Ordóñez, S., Marín, J., Castro, F., & Martínez, V. (2007). Study and methods of analysis of mechanically alloyed Cu–Mo powders. *Materials Science and Engineering: A, 464*(1–2), 288–294. https://doi.org/10.1016/j.msea.2007.02.017.
47. Užarević, K., Halasz, I., & Friščić, T. (2015). Real-time and in situ monitoring of mechanochemical reactions: A new playground for all chemists. *Journal of Physical Chemistry Letters, 6*(20), 4129–4140. https://doi.org/10.1021/acs.jpclett.5b01837.
48. Zhuang, S., Lee, E. S., Lei, L., Nunna, B. B., Kuang, L., & Zhang, W. (2016). Synthesis of nitrogen-doped graphene catalyst by high-energy wet ball milling for electrochemical systems. *International Journal of Energy Research, 40*(15), 2136–2149. https://doi.org/10.1002/er.3595.
49. Peterson, S. C., Jackson, M. A., Kim, S., & Palmquist, D. E. (2012). Increasing biochar surface area: Optimization of ball milling parameters. *Powder Technology, 228,* 115–120. https://doi.org/10.1016/j.powtec.2012.05.005.
50. Ramanayaka, S., Tsang, D. C. W., Hou, D., Ok, Y. S., & Vithanage, M. (2020). Green synthesis of graphitic nano biochar for the removal of emerging contaminants in aqueous media. *Science of The Total Environment, 706,* Article 135725. https://doi.org/10.1016/j.scitotenv.2019.135725.
51. Nath, B. K., Chaliha, C., & Kalita, E. (2019). Iron oxide permeated mesoporous rice-husk nano biochar (IPMN) mediated removal of dissolved arsenic (As): Chemometric modeling and adsorption dynamics. *Journal of Environmental Management, 246,* 397–409. https://doi.org/10.1016/j.jenvman.2019.06.008.
52. Li, W., Zhu, X., He, Y., Xing, B., Xu, J., & Brookes, P. C. (2013). Enhancement of water solubility and mobility of phenanthrene by natural soil nanoparticles. *Environmental Pollution, 176,* 228–233. https://doi.org/10.1016/j.envpol.2013.01.011.
53. Sajjadi, B., Chen, W. Y., Mattern, D. L., Hammer, N., & Dorris, A. (2020). Low-temperature acoustic-based activation of biochar for enhanced removal of heavy metals. *Journal of Water Process Engineering, 34,* Article 101166. https://doi.org/10.1016/j.jwpe.2020.101166.
54. Shan, Y., Yang, W., Li, Y., Liu, Y., & Pan, J. (2019). Preparation of microwave-activated magnetic bio-char adsorbent and study on removal of elemental mercury

from flue gas. *Science of The Total Environment, 697,* Article 134049. https://doi.org/10.1016/j.scitotenv.2019.134049.

55. Yang, Y., Zhou, B., Hu, Z., & Lin, H. (2020). The effects of nano-biochar on maize growth in northern Shaanxi province on the loess plateau. *Applied Ecology and Environmental Research, 18*(2), 2863–2877. https://doi.org/10.15666/aeer/1802_28632877.
56. Anupama, & Khare, P. (2021). A comprehensive evaluation of inherent properties and applications of nano-biochar prepared from different methods and feedstocks. *Journal of Cleaner Production, 320,* Article 128759. https://doi.org/10.1016/j.jclepro.2021.128759.
57. Kanaoujiya, R., Saroj, S. K., Rajput, V. D., Alimuddin, Srivastava, S., Minkina, T., ... & Kumar, A. (2023). Emerging applications of nanotechnology for mankind. *Emergent Materials, 6*(2), 439–452.
58. Ullmann, C., Babick, F., Koeber, R., & Stintz, M. (2017). Performance of analytical centrifugation for the particle size analysis of real-world materials. *Powder Technology, 319,* 261–270. https://doi.org/10.1016/j.powtec.2017.06.057.
59. Song, B., Chen, M., Zhao, L., Qiu, H., & Cao, X. (2019). Physicochemical property and colloidal stability of micron- and nano-particle biochar derived from a variety of feedstock sources. *Science of The Total Environment, 661,* 685–695. https://doi.org/10.1016/j.scitotenv.2019.01.193.
60. Lian, F., et al. (2020). Size matters: Nano-biochar triggers decomposition and transformation inhibition of antibiotic resistance genes in aqueous environments. *Environmental Science & Technology, 54*(14), 8821–8829. https://doi.org/10.1021/acs.est.0c02227.
61. Yue, L., Lian, F., Han, Y., Bao, Q., Wang, Z., & Xing, B. (2019). The effect of biochar nanoparticles on rice plant growth and the uptake of heavy metals: Implications for agronomic benefits and potential risk. *Science of The Total Environment, 656,* 9–18. https://doi.org/10.1016/j.scitotenv.2018.11.364.
62. Xiao, J., Hu, R., & Chen, G. (2020). Micro-nano-engineered nitrogenous bone biochar was developed with a ball-milling technique for high-efficiency removal of aquatic Cd (II), Cu (II), and Pb (II). *Journal of Hazardous Materials, 387,* Article 121980. https://doi.org/10.1016/j.jhazmat.2019.121980.
63. Plácido, J., Bustamante-López, S., Meissner, K. E., Kelly, D. E., & Kelly, S. L. (2019). Nano Refinery of carbonaceous nanomaterials: Complementing dairy manure gasification and their applications in cellular imaging and heavy metal sensing. *Science of The Total Environment, 689,* 10–20. https://doi.org/10.1016/j.scitotenv.2019.06.390.
64. Dong, X., He, L., Liu, Y., & Piao, Y. (2018). Preparation of highly conductive biochar nanoparticles for rapid and sensitive detection of 17β-estradiol in water. *Electrochimica Acta, 292,* 55–62. https://doi.org/10.1016/j.electacta.2018.09.129.
65. Guo, F., et al. (2020). A simple method for the synthesis of biochar nanodots using a hydrothermal reactor. *MethodsX, 7,* Article 101022. https://doi.org/10.1016/j.mex.2020.101022.
66. Ramanayaka, S., Vithanage, M., Alessi, D. S., Liu, W. J., Jayasundera, A. C. A., & Ok, Y. S. (2020). Nano biochar: Production, properties, and multifunctional applications.

Environmental Science: Nano, 7(11), 3279–3302. https://doi.org/10.1039/D0EN00486C.

67. Kumar, M., et al. (2021). A critical review on biochar for enhancing biogas production from anaerobic digestion of food waste and sludge. *Journal of Cleaner Production, 305,* Article 127143. https://doi.org/10.1016/j.jclepro.2021.127143.
68. Sharma, H. C., Chen, H., & Chandel, A. K. (2022). Biochar modification with hematite and goethite as efficient persulfate activation catalysts for sulfamethoxazole degradation: One-step biochar synthesis method and solution matrix effect on sulfamethoxazole removal kinetics. *Environmental Science and Pollution Research, 29*(55), 83931–83944. https://doi.org/10.1007/s11356-022-21743-4.
69. Mohan, D., Kumar, S., & Srivastava, A. (2014). Fluoride removal from groundwater using magnetic and nonmagnetic corn stover biochars. *Ecological Engineering, 73,* 798–808. https://doi.org/10.1016/j.ecoleng.2014.08.017.
70. Wu, J., et al. (2020). A novel calcium-based magnetic biochar is effective in the stabilization of arsenic and cadmium co-contamination in aerobic soils. *Journal of Hazardous Materials, 387,* Article 122010. https://doi.org/10.1016/j.jhazmat.2019.122010.
71. Wei, X., Wang, X., Gao, B., Zou, W., & Dong, L. (2020). Facile ball-milling synthesis of CuO/biochar nanocomposites for efficient removal of Reactive Red 120. *ACS Omega, 5*(11), 5748–5755. https://doi.org/10.1021/acsomega.9b03787.
72. Wang, S., et al. (2015). Removal of arsenic by magnetic biochar prepared from pinewood and natural hematite. *Bioresource Technology, 175,* 391–395. https://doi.org/10.1016/j.biortech.2014.10.104.
73. Li, C., Zhang, L., Gao, Y., & Li, A. (2018). Facile synthesis of nano ZnO/ZnS modified biochar by directly pyrolyzing zinc-contaminated corn stover for Pb (II), Cu (II), and Cr (VI) removals. *Waste Management, 79,* 625–637. https://doi.org/10.1016/j.wasman.2018.08.035.
74. Liu, S. B., et al. (2016). Production of biochars from Ca-impregnated ramie biomass (*Boehmeria nivea* (L.) Gaud.) and their phosphate removal potential. *RSC Advances, 6*(7), 5871–5880. https://doi.org/10.1039/c5ra22142k.
75. Rallet, D., Paltahe, A., Tsamo, C., & Loura, B. (2022). Synthesis of clay-biochar composite for glyphosate removal from aqueous solution. *Heliyon, 8*(3), Article e09112. https://doi.org/10.1016/j.heliyon.2022.e09112.
76. Yao, Y., et al. (2014). Characterization and environmental applications of clay-biochar composites. *Chemical Engineering Journal, 242,* 136–143. https://doi.org/10.1016/j.cej.2013.12.062.
77. Huang, X., Bai, J., Li, K., Zhao, Y., Tian, W., & Hu, C. (2020). Preparation of clay/biochar composite adsorption particle and performance for ammonia nitrogen removal from aqueous solution. *Journal of Ocean University of China, 19*(3), 729–739. https://doi.org/10.1007/s11802-020-4150-9.
78. Li, L., et al. (2017). Mass preparation of micro/nano-powders of biochar with water-dispersibility and their potential application. *New Journal of Chemistry, 41*(18), 9649–9657. https://doi.org/10.1039/c7nj00742f.
79. Hall, K. E., Spokas, K. A., Gamiz, B., Cox, L., Papiernik, S. K., & Koskinen, W. C. (2018). Glyphosate sorption/desorption on biochars – interactions of physical

and chemical processes. *Pest Management Science, 74*(5), 1206–1212. https://doi.org/10.1002/ps.4530.

80. Wallace, C. A., Afzal, M. T., & Saha, G. C. (2019). Effect of feedstock and microwave pyrolysis temperature on physio-chemical and nano-scale mechanical properties of biochar. *Bioresources and Bioprocessing, 6*(1), Article 268. https://doi.org/10.1186/s40643-019-0268-2.
81. Luo, J., et al. (2018). Sorption of norfloxacin, sulfamerazine, and oxytetracycline by KOH-modified biochar under single and ternary systems. *Bioresource Technology, 263,* 385–392. https://doi.org/10.1016/j.biortech.2018.05.022.
82. Omoriyekomwan, J. E., Tahmasebi, A., Zhang, J., & Yu, J. (2017). Formation of hollow carbon nanofibers on bio-char during microwave pyrolysis of palm kernel shell. *Energy Conversion and Management, 148,* 583–592. https://doi.org/10.1016/j.enconman.2017.06.022.
83. Oleszczuk, P., Ćwikła-Bundyra, W., Bogusz, A., Skwarek, E., & Ok, Y. S. (2016). Characterization of nanoparticles of biochars from different biomass. *Journal of Analytical and Applied Pyrolysis, 121,* 165–172. https://doi.org/10.1016/j.jaap.2016.07.017.
84. Huang, H. J., Yang, T., Lai, F. Y., & Wu, G. Q. (2017). Co-pyrolysis of sewage sludge and sawdust/rice straw for the production of biochar. *Journal of Analytical and Applied Pyrolysis, 125,* 61–68. https://doi.org/10.1016/j.jaap.2017.04.018.
85. Fan, L., et al. (2016). Chelating capture and magnetic removal of non-magnetic heavy metal substances from soil. *Scientific Reports, 6,* Article 21027. https://doi.org/10.1038/srep21027.
86. Zhao, B., Xu, X., Zeng, F., Li, H., & Chen, X. (2018). The hierarchical porous structure bio-char assessments produced by co-pyrolysis of municipal sewage sludge and hazelnut shell and Cu (II) adsorption kinetics. *Environmental Science and Pollution Research, 25*(20), 19423–19435. https://doi.org/10.1007/s11356-018-2079-y.
87. Plácido, J., Bustamante López, S., Meissner, K. E., Kelly, D. E., & Kelly, S. L. (2019). Multivariate analysis of biochar-derived carbonaceous nanomaterials for detection of heavy metal ions in aqueous systems. *Science of The Total Environment, 688,* 751–761. https://doi.org/10.1016/j.scitotenv.2019.06.342.
88. Wang, Y., & Liu, R. (2017). Comparison of characteristics of twenty-one types of biochar and their ability to remove multi-heavy metals and methylene blue in solution. *Fuel Processing Technology, 160,* 55–63. https://doi.org/10.1016/j.fuproc.2017.02.019.
89. Meng, J., Tao, M., Wang, L., Liu, X., & Xu, J. (2018). Changes in heavy metal bioavailability and speciation from a Pb-Zn mining soil amended with biochars from co-pyrolysis of rice straw and swine manure. *Science of The Total Environment, 633,* 300–307. https://doi.org/10.1016/j.scitotenv.2018.03.199.
90. Zheng, H., et al. (2013). Characteristics and nutrient values of biochars produced from giant reed at different temperatures. *Bioresource Technology, 130,* 463–471. https://doi.org/10.1016/j.biortech.2012.12.044.
91. Nartey, O. D., & Zhao, B. (2014). Biochar preparation, characterization, and adsorptive capacity and its effect on bioavailability of contaminants: An overview. *Advances in Materials Science and Engineering, 2014,* Article 715398. https://doi.org/10.1155/2014/715398.

92. Ghani, W. A. W. A. K., et al. (2013). Biochar production from waste rubber-wood-sawdust and its potential use in C sequestration: Chemical and physical characterization. *Industrial Crops and Products, 44,* 18–24. https://doi.org/10.1016/j.indcrop.2012.10.017.
93. Kasozi, G. N., Zimmerman, A. R., Nkedi-Kizza, P., & Gao, B. (2010). Catechol and humic acid sorption onto a range of laboratory-produced black carbons (biochars). *Environmental Science and Technology, 44*(16), 6189–6195. https://doi.org/10.1021/es1014423.
94. Luo, C., Lü, F., Shao, L., & He, P. (2015). Application of eco-compatible biochar in anaerobic digestion to relieve acid stress and promote the selective colonization of functional microbes. *Water Research, 68,* 710–718. https://doi.org/10.1016/j.watres.2014.10.052.
95. Xu, X., Zheng, Y., Gao, B., & Cao, X. (2019). N-doped biochar synthesized by a facile ball-milling method for enhanced sorption of CO_2 and reactive red. *Chemical Engineering Journal, 368,* 564–572. https://doi.org/10.1016/j.cej.2019.02.165.
96. Huang, J., Zimmerman, A. R., Chen, H., & Gao, B. (2020). Ball-milled biochar effectively removes sulfamethoxazole and sulfapyridine antibiotics from water and wastewater. *Environmental Pollution, 258,* Article 113809. https://doi.org/10.1016/j.envpol.2019.113809.
97. Chen, B., & Chen, Z. (2009). Sorption of naphthalene and 1-naphthol by biochars of orange peels with different pyrolytic temperatures. *Chemosphere, 76*(1), 127–133 https://doi.org/10.1016/j.chemosphere.2009.02.004.
98. Rafiq, M. K., Bachmann, R. T., Rafiq, M. T., Shang, Z., Joseph, S., & Long, R. L. (2016). Influence of pyrolysis temperature on physicochemical properties of corn stover (*Zea mays* L.) biochar and feasibility for carbon capture and energy balance. *PLOS One, 11*(6), e0156894. https://doi.org/10.1371/journal.pone.0156894.
99. Lyu, H., et al. (2018). Experimental and modeling investigations of ball-milled biochar for the removal of aqueous methylene blue. *Chemical Engineering Journal, 335,* 110–119. https://doi.org/10.1016/j.cej.2017.10.130.
100. Zheng, Y., Wan, Y., Chen, J., Chen, H., & Gao, B. (2020). MgO modified biochar produced through ball milling: A dual-functional adsorbent for removal of different contaminants. *Chemosphere, 243,* Article 125344. https://doi.org/10.1016/j.chemosphere.2019.125344.
101. Sethupathi, S., et al. (2017). Biochars as potential adsorbers of CH_4, CO_2, and H_2S. *Sustainability, 9*(1), 121. https://doi.org/10.3390/su9010121.
102. Fiore, S., Berruti, F., & Briens, C. (2018). Investigation of innovative and conventional pyrolysis of ligneous and herbaceous biomasses for biochar production. *Biomass and Bioenergy, 119,* 381–391. https://doi.org/10.1016/j.biombioe.2018.10.010.
103. Chen, T., et al. (2014). Influence of pyrolysis temperature on characteristics and heavy metal adsorptive performance of biochar derived from municipal sewage sludge. *Bioresource Technology, 164,* 47–54. https://doi.org/10.1016/j.biortech.2014.04.048.
104. Harmsen, J., & Naidu, R. (2013). Bioavailability as a tool in site management. *Journal of Hazardous Materials, 261,* 840–846. Available at. https://www.academia.edu/16587921/Bioavailability_as_a_tool_in_site_management (accessed on 19 June 2024).

105. Murage, E. W., Voroney, P., & Beyaert, R. P. (2007). Turnover of carbon in the free light fraction with and without charcoal as determined using the $_{13}$C natural abundance method. *Geoderma, 138*(1–2), 133–143. https://doi.org/10.1016/j.geoderma.2006.11.002.
106. Haefele, S. M., et al. (2011). Effects and fate of biochar from rice residues in rice-based systems. *Field Crops Research, 121*(3), 430–440. https://doi.org/10.1016/j.fcr.2011.01.014.
107. Schimmelpfennig, S., & Glaser, B. (2012). One step forward toward characterization: Some important material properties to distinguish biochars. *Journal of Environmental Quality, 41*(4), 1001–1013. https://doi.org/10.2134/jeq2011.0146.
108. Hammes, K., Smernik, R., Skjemstad, J., & Schmidt, M. (2008). Characterization and evaluation of reference materials for black carbon analysis using elemental composition, color, BET surface area, and $_{13}$C NMR spectroscopy. *Organic Geochemistry, 39*(8), 1206–1222. https://doi.org/10.1016/j.apgeochem.2008.04.023.
109. Yu, J. T., Dehkhoda, A. M., & Ellis, N. (2011). Development of biochar-based catalyst for transesterification of canola oil. *Energy and Fuels, 25*(1), 337–344. https://doi.org/10.1021/ef100977d.
110. Cheah, S., Malone, S. C., & Feik, C. J. (2014). Speciation of sulfur in biochar produced from pyrolysis and gasification of oak and corn stover. *Environmental Science and Technology, 48*(15), 8474–8480. https://doi.org/10.1021/es500073r.
111. Qu, X., Fu, H., Mao, J., Ran, Y., Zhang, D., & Zhu, D. (2016). Chemical and structural properties of dissolved black carbon released from biochars. *Carbon, 96,* 759–767. https://doi.org/10.1016/j.carbon.2015.09.106.
112. Zhao, L., Cao, X., Mašek, O., & Zimmerman, A. (2013). Heterogeneity of biochar properties as a function of feedstock sources and production temperatures. *Journal of Hazardous Materials, 256–257,* 1–9. https://doi.org/10.1016/j.jhazmat.2013.04.015.
113. Kanjanarong, J., et al. (2017). Removal of hydrogen sulfide generated during anaerobic treatment of sulfate-laden wastewater using biochar: Evaluation of efficiency and mechanisms. *Bioresource Technology, 234,* 115–121. https://doi.org/10.1016/j.biortech.2017.03.009.
114. Inyang, M., Gao, B., Pullammanappallil, P., Ding, W., & Zimmerman, A. R. (2010). Biochar from anaerobically digested sugarcane bagasse. *Bioresource Technology, 101*(22), 8868–8872. https://doi.org/10.1016/j.biortech.2010.06.088.
115. Liang, B., et al. (2006). Black carbon increases cation exchange capacity in soils. *Soil Science Society of America Journal, 70*(5), 1719–1730. https://doi.org/10.2136/sssaj2005.0383.
116. Chacón, F. J., Sánchez-Monedero, M. A., Lezama, L., & Cayuela, M. L. (2020). Enhancing biochar redox properties through feedstock selection, metal preloading, and post-pyrolysis treatments. *Chemical Engineering Journal, 395,* Article 125100. https://doi.org/10.1016/j.cej.2020.125100.
117. Klüpfel, L., Keiluweit, M., Kleber, M., & Sander, M. (2014). Redox properties of plant biomass-derived black carbon (biochar). *Environmental Science and Technology, 48*(10), 5601–5611. https://doi.org/10.1021/es500906d.
118. Dieguez-Alonso, A., et al. (2019). Designing biochar properties through the blending of biomass feedstock with metals: Impact on oxyanions adsorption behavior. *Chemosphere, 214,* 743–753. https://doi.org/10.1016/j.chemosphere.2018.09.091.

119. Dixit, C. (2012). Surface modification and conjugation strategies for bioassay/biomaterial applications (PhD thesis), 1–24. Dublin City University.
120. Huang, D., et al. (2017). Sorptive removal of ionizable antibiotic sulfamethazine from aqueous solution by graphene oxide-coated biochar nanocomposites: Influencing factors and mechanism. *Chemosphere, 186,* 414–421. https://doi.org/10.1016/j.chemosphere.2017.07.154.
121. Goswami, L., et al. (2022). Nano-Biochar as a sustainable catalyst for anaerobic digestion: A synergetic closed-loop approach. *Catalysts, 12*(2), 186. https://doi.org/10.3390/catal12020186.
122. Patel, S., et al. (2020). Production of hydrogen by catalytic methane decomposition using biochar and activated char produced from biosolids pyrolysis. *International Journal of Hydrogen Energy, 45*(55), 29978–29992. https://doi.org/10.1016/j.ijhydene.2020.08.036.
123. Sugiarto, Y., Sunyoto, N. M. S., Zhu, M., Jones, I., & Zhang, D. (2021). Effect of biochar in enhancing hydrogen production by mesophilic anaerobic digestion of food wastes: The role of minerals. *International Journal of Hydrogen Energy, 46*(5), 3695–3703. https://doi.org/10.1016/j.ijhydene.2020.10.256.
124. Zhang, D., et al. (2017). Aerobic granulation accelerated by biochar for the treatment of refractory wastewater. *Chemical Engineering Journal, 314,* 88–97. https://doi.org/10.1016/j.cej.2016.12.128.
125. Ema, M., Hougaard, K. S., Kishimoto, A., & Honda, K. (2015). Reproductive and developmental toxicity of carbon-based nanomaterials: A literature review. *Nanotoxicology, 10*(4), 391–412. https://doi.org/10.3109/17435390.2015.1073811.
126. Freixa, A., Acuña, V., Sanchís, J., Farré, M., Barceló, D., & Sabater, S. (2018). Ecotoxicological effects of carbon-based nanomaterials in aquatic organisms. *Science of The Total Environment, 619–620,* 328–337. https://doi.org/10.1016/j.scitotenv.2017.11.095.
127. Chen, R., et al. (2010). Differential uptake of carbon nanoparticles by plant and mammalian cells. *Small, 6*(5), 612–617. https://doi.org/10.1002/smll.200901911.
128. Liu, X., Tang, J., Wang, L., Liu, Q., & Liu, R. (2019). A comparative analysis of ball-milled biochar, graphene oxide, and multi-walled carbon nanotubes with respect to toxicity induction in Streptomyces. *Journal of Environmental Management, 243,* 308–317. https://doi.org/10.1016/j.jenvman.2019.05.034.
129. Ndirangu, S. M., Liu, Y., Xu, K., & Song, S. (2019). Risk evaluation of pyrolyzed biochar from multiple wastes. *Journal of Chemistry, 2019,* 4506314. https://doi.org/10.1155/2019/4506314.
130. Kushwaha, A., et al. (2022). Anaerobic digestion as a sustainable biorefinery concept for waste-to-energy conversion. In *Waste-to-Energy Approaches Towards Zero Waste: Interdisciplinary Methods of Controlling Waste* (pp. 129–163). https://doi.org/10.1016/B978-0-323-85387-3.00008-2.
131. Song, B., et al. (2022). Preparation of nano-biochar from conventional biorefineries for high-value applications. *Renewable and Sustainable Energy Reviews, 157,* 112057. https://doi.org/10.1016/j.rser.2021.112057.
132. Rajput, V. D., Minkina, T., Kumari, A., Harish, Singh, V. K., Verma, K. K., Mandzhieva, S., Sushkova, S., Srivastava, S., & Keswani, C. (2021). Coping with

the challenges of abiotic stress in plants: New dimensions in the field application of nanoparticles. *Plants, 10*(6), 1221. https://doi.org/10.3390/plants10061221.

133. Rajput, V., Minkina, T., Sushkova, S., Behal, A., Maksimov, A., Blicharska, E., Ghazaryan, K., Movsesyan, H., & Barsova, N. (2019). ZnO and CuO nanoparticles: A threat to soil organisms, plants, and human health. *Environmental Geochemistry and Health, 42*(1), 147–158. https://doi.org/10.1007/s10653-019-00317-3.
134. Rajput, V. D., Minkina, T., Sushkova, S., Tsitsuashvili, V., Mandzhieva, S., Gorovtsov, A., Nevidomskyaya, D., & Gromakova, N. (2017). Effect of nanoparticles on crops and soil microbial communities. *Journal of Soils and Sediments, 18*(6), 2179–2187. https://doi.org/10.1007/s11368-017-1793-2.
135. Rajput, V., Minkina, T., Fedorenko, A., Sushkova, S., Mandzhieva, S., Lysenko, V., Duplii, N., Fedorenko, G., Dvadnenko, K., & Ghazaryan, K. (2018). Toxicity of copper oxide nanoparticles on spring barley (*Hordeum sativum* Distichum). *Science of The Total Environment, 645,* 1103–1113. https://doi.org/10.1016/j.scitotenv.2018.07.211.
136. Rajput, V. D., Kumari, A., Minkina, T., Barakhov, A., Singh, S., Mandzhieva, S. S., Sushkova, S., Ranjan, A., Rajput, P., & Garg, M. C. (2022). A practical evaluation of the integrated role of biochar and nanomaterials in soil remediation processes. *Environmental Geochemistry and Health, 45, 9435–9449.* https://doi.org/10.1007/s10653-022-01375-w.
137. Rajput, V. D., Faizan, M., Upadhyay, S. K., Kumari, A., Ranjan, A., Sushkova, S., Chauhan, P. K., Mahmud, A. A., Burachevskaya, M., Chaplygin, V., & Deryabkina, I. (2022). Influence of nanoparticles on the plant rhizosphere microbiome. In *Sustainable Plant Nutrition in a Changing World* (pp. 83–102). https://doi.org/10.1007/978-3-030-97389-6_4.
138. Rajput, V. D., Singh, A., Minkina, T., Rawat, S., Mandzhieva, S., Sushkova, S., Shuvaeva, V., Nazarenko, O., Rajput, P., Komariah, Verma, K. K., Singh, A. K., Rao, M., & Upadhyay, S. K. (2021). Nano-enabled products: Challenges and opportunities for sustainable agriculture. *Plants, 10*(12), 2727. https://doi.org/10.3390/plants10122727.
139. Singh, A., Singh, A. K., Rawat, S., Pal, N., Rajput, V. D., Minkina, T., Sharma, R., Singh, N. P., & Tripathi, J. N. (2022). Satellite-based quantification of methane emissions from wetlands and rice paddies ecosystems in North and Northeast India. *Hydrobiology, 1*(3), 317–330. https://doi.org/10.3390/hydrobiology1030023.
140. Singh, A., Sengar, R. S., Rajput, V. D., Minkina, T., & Singh, R. K. (2022). Zinc oxide nanoparticles improve salt tolerance in rice seedlings by improving physiological and biochemical indices. *Agriculture, 12*(7), 1014. https://doi.org/10.3390/agriculture12071014.
141. Singh, A., Sengar, R. S., Sharma, R., & Rajput, P. (2021). Nano-priming technology for sustainable agriculture. *Biogeosystem Technique, 8*(2), 79–92. https://doi.org/10.13187/bgt.2021.2.79.
142. Singh, A., Mehrotra, R., Rajput, V. D., Dmitriev, P., Singh, A. K., Kumar, P., Tomar, R. S., Singh, O., & Singh, A. K. (2022). Geoinformatics, artificial intelligence, sensor technology, big data. In *Sustainable Agriculture Systems and Technologies* (pp. 295–313). https://doi.org/10.1002/9781119808565.ch14.
143. Singh, A., Rajput, V. D., Rawat, S., Sharma, R., Singh, A. K., Kumar, P., Singh, A. K., Minkina, T., Singh, R. P., & Singh, S. (2022). Geoinformatics and

nanotechnological approaches for coping up abiotic and biotic stress in crop plants. In *Sustainable Agriculture Systems and Technologies* (pp. 337–359). https://doi.org/10.1002/9781119808565.ch17.

144. Singh, A., Rajput, V. D., Rawat, S., Sharma, R., Kumar Singh, A., Kumar Singh, A., & Tomar, R. S. (2021). Emerging tools for sustainable agriculture and food security: An introduction. In *Emerging Tools for Sustainable Agriculture and Food Security* (1st ed., pp. 3–16). Deepika: New Delhi.
145. Rajput, P., Rajput, V. D., Singh, A., Sharma, R., Rawat, S., Tomar, R. S., Singh, A., & Minkina, T. (2021). Impact of emerging tools for solving 'hidden hunger' in humans. *Emerging Tools for Sustainable Agriculture and Food Security, 1*(1), 17–30.
146. Rajput, V. D., Minkina, T., Kumari, A., Shende, S. S., Ranjan, A., Faizan, M., Barakvov, A., Gromovik, A., Gorbunova, N., Rajput, P., Singh, A., Khabirov, I., Nazarenko, O., Sushkova, S., & Kizilkaya, R. (2022). A review on nano bioremediation approaches for restoration of contaminated soil. *Eurasian Journal of Soil Science, 11*(1), 43–60. https://doi.org/10.18393/ejss.990605.
147. Rajput, V. D., Minkina, T., Feizi, M., Kumari, A., Khan, M., Mandzhieva, S., Sushkova, S., El-Ramady, H., Verma, K. K., Singh, A., Hullebusch, E. D. van, Singh, R. K., Jatav, H. S., & Choudhary, R. (2021). Effects of silicon and silicon-based nanoparticles on rhizosphere microbiome, plant stress and growth. *Biology, 10*(8), 791. https://doi.org/10.3390/biology10080791.
148. Singh, A., et al. (2020). Monitoring soil salinity and recent advances in the mechanism of salinity tolerance in plants. *Bio geosystem Technique, 7*(2), 66–87. https://doi.org/10.13187/bgt.2020.2.66.
149. Singh, A., Sengar, R. S., Shahi, U. P., Rajput, V. D., Minkina, T., & Ghazaryan, K. A. (2022). Prominent effects of zinc oxide nanoparticles on roots of rice (*Oryza sativa* L.) grown under salinity stress. *Stresses, 3*, 33–46. https://doi.org/10.3390/stresses3010004.

CHAPTER 2

Application of Nano-Biochar to Improve Soil Quality and Sustainability

SHRENI AGRAWAL,[1] PRADEEP KUMAR,[2] RICHA DAS,[1]
AMIT KUMAR SINGH,[3] PRAVEEN KUMAR SHUKLA,[2] POOJA VERMA,[2]
VISHNU D. RAJPUT,[4] INDRANI BHATTACHARYA,[1]
SUNIL KUMAR MISHRA,[3] and KAVINDRA NATH TIWARI[2]

[1]*Department of Biotechnology, Parul Institute of Applied Science, Parul University, Vadodara, Gujarat, India*

[2]*Department of Botany, MMV, Banaras Hindu University, Varanasi, Uttar Pradesh, India*

[3]*Department of Pharmaceutical Engineering and Technology, Indian Institute of Technology, Banaras Hindu University, Varanasi, Uttar Pradesh, India*

[4]*Academy of Biology and Biotechnology, Southern Federal University, Rostov on Don, Russia*

ABSTRACT

The Green Revolution has been beneficial in promoting the growth of human civilization, but it has also degraded the soil, destroyed biodiversity, and accelerated climate change. Advanced nanomaterials, including nano-biochar, have provided prolonged solutions for a wide range of current challenges. Nano-biochar is a specialized form of biochar with a structural size on the nanometer scale featuring better morphological and physiochemical properties. Nano-BC application improves soil qualities, making it better suited for plant growth and development. By enhancing soil

Nano-Biochar and Nanoparticles for Soil Health Improvement.
Vishnu D. Rajput and Tatiana M. Minkina (Eds.)

porosity, resistivity, and water-holding capacities—all crucial for sustaining soil activity—Nano-BC offers an ideal soil habitat for bacteria. Biomass is pyrolyzed to produce the bulk parent biochar, which is then mechanically processed using various milling methods to generate nano-biochar. Different types of nano-biochar, such as biochar nanocomposites, magnetic nano-biochar, functional nanoparticles coated nano-biochar, and colloidal biochar, have greater environmental applications than normal biochar, such as improving plant growth, removing pesticides from soil, adding fertilizer, microbial growth, and managing disease. A large surface area, high degree of crystallinity, high nutritional quality, and good chemical group concentrations are a few of the distinctive characteristics of nano-biochar. The features of biochar basically depend on biomaterials that were utilized and the pyrolysis circumstances, with its distinctiveness resting on its large specific surface area and a constant source of carbon, which predispose to superior crop responses and soil health. These chapters discuss the production, types, and various applications of nano-biochar, as well as their significant contributions to agriculture, particularly in soil development.

2.1 INTRODUCTION

Soil is a precious ecological component that promotes plant growth by supplying water, minerals, and structural support. It is the base of crop production in agriculture. Nowadays, constant soil modification occurs in an agricultural ecosystem largely by feeding, nutrient removal, altering water balance, and bacterial life. These activities have an influence on the mechanical, and biochemical features of the soil, and the degree by which many properties deviate from the norm being determined by the topsoil neutralizing potential as well as tolerance. If such alterations exceed the capacity of soil resistance, soil loses its physical integrity, resulting in soil deterioration [1].

Increasing commercialization, globalization, and demographic change also affect the ecosystem and cause soil erosion. Additionally, it leads to the production of a large amount of biowaste, such as agricultural waste, municipal solid waste (MSW), commercial waste, marine waste, forestry garbage, etc. [2]. Scientists have developed efficient regulatory practices and strategies, such as the use of mulching, cover crops, minimizing tillage, contour bonds, planting trees, wood logs, and plowing across slopes, in an effort to control soil degradation off-site and on-site and to improve soil production in the agriculture field [3, 4]. One method to address these issues is the pyrolysis of biomass (organic waste) to create biochar, which is subsequently administered to agricultural land [5].

Biochar, also referred to as charcoaled biomass or black carbon, is an organic substance produced through the pyrolysis of various types of carbon-based material, including agricultural residue, waste products of animals, poultry, as well as metropolitan waste, at high temperatures (300 to 1,000°C) and under partial or anaerobic conditions (Figure 2.1) [6, 7]. Because of its abilities in agricultural productivity, wastewater treatment, carbon sequestration, bioremediation, and significant environmental management solutions in agriculture, biochar has drawn a lot of interest in the past 20 years [6]. Its extremely porous nature allows for the presence of exportable levels of humic and fulvic-like compounds [8]. Additionally, the molecular composition of the substance exhibits excellent microbiological and chemical resilience [9]. Several factors like pyrolysis temperature, holding time, burner temperature, and biomass types play a significant effect on the physiochemical characteristics of biochar [10, 11]. In contrast to non-pyrolyzed organic compounds like plant wastes and agricultural residues, biochar is resistant to microbial degradation due to its refractory aromatic carbon structure, which enhances soil organic carbon (SOC) by remaining in the soil for up to 100–1,000 years [12].

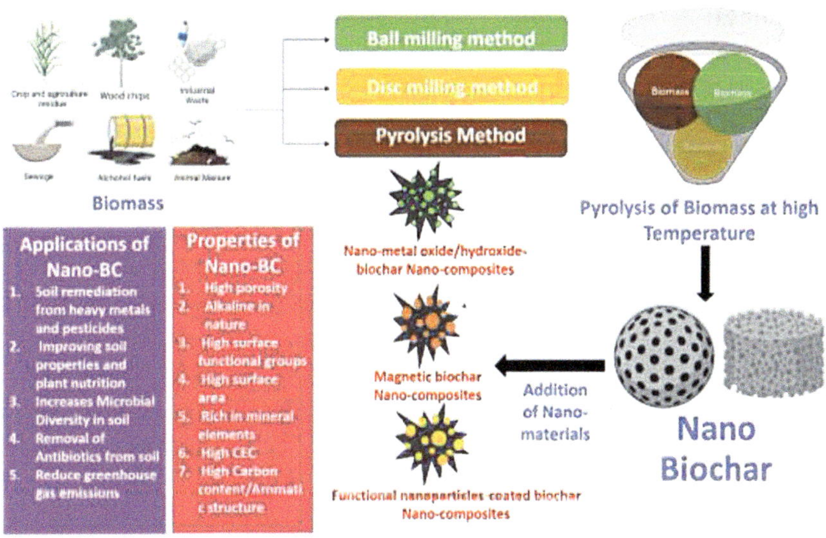

FIGURE 2.1 Graphical abstract representing the fabrication methods, types, properties, and applications of nano biochar.

The utilization of biochar as a natural fertilizer has a variety of potential advantages, including improved soil nutrient supply and bioaccumulation [13, 14], increased effectiveness of using inorganic and organic fertilizers [15], enhanced uptake of soil mineral N (NH_4 and NO_3) [16], decreased nitrogen leakage and the requirement for fertilizer [17], increased soil qualities (soil permeability, relative density, porosity, bulk stability, infiltration capacity, and hydrological conductance) [18, 19], Enhanced base absorption and liming effects in acidic soils [20], decrease the availability of toxic metal ions to plants growing in polluted soils and increase their stability [21, 22] encouraging the growth and activity of microbial communities [23], enhanced agricultural yields and sustainability [24], improved symbiotic nitrogen absorption in legumes [25], reduced salinity, dryness, and thermal stress on crop development and soil conditions [26–28]. Furthermore, adding biochar to soils is a low-cost way for stabilizing sustainably grown compounds as well as lowering emissions of greenhouse gases [29].

The increasing use of biochar has changed agricultural yield and opened up new opportunities for environmental restoration. Bulk biochar (B-BCs, 0.04–20 mm) is often used for environmental and agricultural stability [30]. The size of biochar particles differs based on the pyrolysis process, ranging from micrometers to centimeters [31]. Reduced particle size enhances the amount of adsorption surfaces existing, improving holding capability [32–34]. Qualities of biochar can be improved by gradually reducing particle size to the nanoscale, or at least 100 nm or smaller. The ratio of surface area to volume is increased, resulting in increased surface tension as well as uptake capacity and, finally, biological productivity [34–37]. Because of its nanoscale properties which favor ecological control for potential applications [38]. By enhancing its physical, chemical, and structural qualities, nano biochar has been developed to go beyond the limitations of biochar. For instance, its nanosize increases surface area, and the abundance of efficient groups by enhanced physical and thermal integrity makes it different and useful [39].

Nano-BC can exhibit a variety of chemical components, aromatic/polar natures, pH, cation transfer capacities, specific surface areas (SSA), porous structure properties, and Zeta potentials [40]. A variety of processes, including mechanical grinding, ultrasonic vibrating, ball milling, double disc milling, and hydrothermal reaction procedures, are used to produce nano-biochar from conventional biochar. According to the current literature, the method of producing nano-biochar by ball milling was deemed to be the best-suited and favored one [41]. The variables used during biomass pyrolysis, such as milling time and temperature, mostly affect the size of BC particles. For instance, by raising the pyrolytic temperature and grinding time, the size of BCs can be increased [38]. Based on the nanomaterials utilized in the manufacturing of nano-biochar, they are divided into three groups: magnetic biochar nanocomposites, oxide/hydroxide biochar nanocomposites, and functional nanoparticles coated biochar mediated via graphene, graphene oxide, chitosan, carbon nanotubes, ZnS nanocrystals, graphitic C_3N_4, etc.

These nanocomposites composed of biochar are incredibly effective in treating a variety of wastes and have a wide range of ecological advantages [42]. It primarily focuses on enhancing plant productivity, managing human disease by eliminating antibiotics from plants, increasing soil restoration in metal-contaminated soil, promoting nutritional absorption in soil, microbial activity, pesticide remediation, lowering carbon sequestration, and fertilizers. Due to its physio-chemical characteristics and porous structure, nano-BC is currently a viable solution for soil alteration. Nano-BC is an efficient supporter of enzymes and a powerful adsorptive material. Nano-BC has specific features like higher surface area, hydrophobicity, and permeability, and also displays higher adsorptive ability for a diversity of toxic pollutants (pesticides, toxic metals, polychlorinated biphenyls (PCBs), and polycyclic aromatic hydrocarbons (PAHs)). They may be used to mitigate soil acidity because of their increased alkalinity and pH [43, 44]. Essential micro- and macronutrients, as well as dangerous pollutants like pesticides and possibly poisonous metals, may be efficiently controlled by nano-BC [38]. Similarly, nano-BC has several applications in the domains of aquaculture, waste management, health care implications, aquatic systems, and microbiological development. In this review paper, more fabrication techniques, nano-BC types, and applications are discussed.

2.2 NANO-BC FABRICATION

The increasing use of nanobiochar has changed agricultural yield and opened up new opportunities for environmental remediation. Nano-BC has more appealing green initiatives for potential uses due to its incredibly small size, variable mobility, solubility, and subsequent impact on soil and vegetation systems. Several of the processes utilized in nano-BC production include ball milling, pyrolysis, and disc milling [38].

2.2.1 BALL MILLING METHOD

In recent years, ball milling has been popular due to its affordability, wide range of applications, and low energy consumption during manufacturing. The material is reduced to the nanoscale via ball milling, which preserves the crystal structure of the material [45]. During the procedure, stainless steel balls as well as agate were utilized. The diameter and weight of the ball also affect the milling process of the material. Milling biomasses included several plant waste materials like bamboo, bagasse of sugarcane, pinewood, rice husk, and wood of Carya. In addition to the type of milling, the rotating speed, ball-to-power mass ratio, and milling time are some of the variables that can directly affect the size and surface energy of the resulting nano-BC particles [46]. Synthesis of nano-BC from milling of pine wood biochar took place using a planetary ball mill [46]. In ball milling methods, a stainless-steel jar and balls were used. To control the overheating of the biochar and balls, the time was set at 5 minutes ON and 5 minutes OFF. Different variables like rotating speed, grinding time, and ball-to-powder ratio were optimized to improve the milling conditions for the production of nano-BC. To achieve the 60 nm size of Nano-BC, the desired ball-to-powder mass ratio (45 g/ball) was standardized.

2.2.2 DISC MILLING METHOD

In another study, nano-BC was prepared via disc milling technique. From the dendrothermal power plant in Thirappane, Sri Lanka, woody biochar was obtained, a byproduct of the gasification of *Gliricidia sepium*. The biochar was chopped into pieces, sieved (<3 mm), rinsed with distilled water, and dried. It was then mechanically ground using a disc mill. A disc mill was used to grind dried biochar in the presence of ethanol for two minutes at a speed of 1,000 rpm. About 5 g of the colloidal biochar was centrifuged at 2,000 rpm for 10 minutes after being mixed with 100 mL of ethanol. The supernatant, which comprised the nano fraction of biochar, was collected, subjected to 30 minutes of sonication at 50 kHz, and then dried in Petri dishes at 50°C using a vacuum oven. A spatula was used to remove the dried layer of nano-BC, and samples were collected for characterization and adsorption tests [48].

2.2.3 PYROLYSIS METHOD

The pyrolytic method is also employed for the nano-BC synthesis. Iron oxide permeated mesoporous rice-husk nano-BC (IPMN) was synthesized using a chemically modified pyrolytic method which can be used in the removal of arsenic. The IPMNs were created using a one-pot pyrolytic method. In this technique, powder of rice husk was mixed with a solution of ferrous sulfate heptahydrate (0.3 M) along with urea (1M). The contents were thoroughly blended with a mortar and pestle to create a fine slurry. For the synthesis of IPMNs, the material was placed into a furnace and pyrolysis took place at 600°C for 5 hours. After the pyrolytic process, IPMNs were kept at room temperature and subjected to consecutive rinsing with acetone, pure alcohol, and distilled water. This was done with the intention of removing any and all remnants of the precursors. IPMNs were then kept in an oven at 60C for 12 hours for complete removal of moisture content [49].

2.3 TYPES OF NANO-BC

The biochar-based nanocomposites can be divided into three groups: functional nanoparticles-coated biochar, magnetic biochar composites, and nano-metal oxide/hydroxide-biochar composites. These biochar nanocomposites are categorized depending on the many types of nanomaterials that are included in their production. Such nanocomposite substances were produced with the help of several distinct biotechnological processes. Every technique of biosynthesis requires at least two processing phases to be completed for the biomass: a pre-treatment step as well as a post-treatment process [50].

2.3.1 METAL OXIDE/HYDROXIDE MEDIATED NANO-BIOCHAR

To construct nanometal oxide/hydroxide-biochar composites, three different fabrication techniques are used (Figure 2.2(A)), such as target element enrichment through bioaccumulation, pre-treating biomass with a metal salt, and adding metal oxide nanoparticles after pyrolysis. The first two techniques are focused on treating biomass by inoculating it with the target metal prior to pyrolysis. In the 3rd method, biochar made from biomass is pyrolyzed directly with nano-metal oxide/hydroxide. By bio-accumulating desired chemical components in biomass, it is possible to make tailored biochar with specific qualities. After being thermally processed, the metal components enhanced in the biomass can be converted into nano-metal oxide/hydroxide [50]. Chemically pre-treating biomass prior to pyrolysis is another strategy. The chemical agents of choice for pre-treating biomass are often metal salts. After the biomass has been immersed in a metal salt solution, the metals may adhere to the surface or enter the interior. Metal ions will change into nano-metal oxide or metal hydroxide after pyrolysis, and biomass that has been saturated with metal ions will turn into biochar-based nanocomposites [51]. Following the pyrolysis of biomass, it is also possible to create biochar-based nanocomposites containing metal oxide nanoparticles adhered to the charcoal matrix [50]. Due to their fine-grained structure, nano-metal oxides have a tendency to agglomerate and passivate despite having a large surface energy and high SSA. Nano-metal oxide-biochar composites (NMOBCs), which incorporate the advantages of both biochar and nano-metal oxides, have become a class of potent adsorbents. The adsorption capacity of both nano-metal oxides and biochar is maximized by NMOBCs, which increase the consistency and dispersity of nano-metal oxides [52].

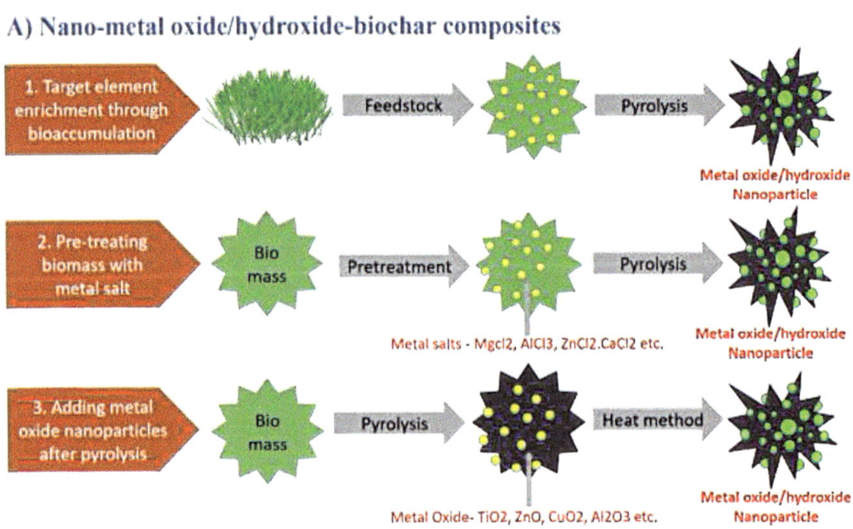

FIGURE 2.2(a). Fabrication methods to construct the metal oxide/hydroxide nanoparticles.

2.3.2 MAGNETIC BIOCHAR COMPOSITES

Given how challenging it is to extract from the aqueous medium, biochar can also be transformed into magnetic materials utilizing two synthetic techniques: pre-treating biomass with iron ions before pyrolysis and chemical co-precipitation of iron oxides onto biochar after pyrolysis (Figure 2.2(B)). These techniques allow for the coating of biochar with nanoscale magnetic iron oxide, introducing magnetization and iron oxide active sites for pollutant removal [50]. The first technique involves chemically co-precipitating or impregnating biomass with Fe_{3+}/Fe_{2+}, then pyrolyzing or warming it in a microwave under various circumstances. Through thermal pyrolysis of $FeCl_3$ pre-treated biomass, a new magnetic biochar-based adsorbent containing colloidal or nanosized-Fe_2O_3 particles incorporated in a porous biochar matrix was created [53]. The final composite displayed excellent ferromagnetic characteristics as well as high adsorption capacity in aqueous solution. One of the most extensively utilized approaches was the chemical co-precipitation of ferric oxide through charcoal after pyrolysis [54–56]. In this technique, biochar was first created by pyrolyzing biomass, then an aqueous ferrous/ferric mixture was combined with an aqueous biochar suspension and finally underwent aqueous NaOH treatment [55, 57]. Baig et al. compared the two synthesis processes for creating magnetic biochar. The findings revealed that the biochar made by chemical co-precipitation of iron oxides onto biomass before pyrolysis had better physical and chemical parameters (Fe_3O_4 loading rate, saturation magnetization, and thermal properties) and adsorption efficacy than the earlier approach [58].

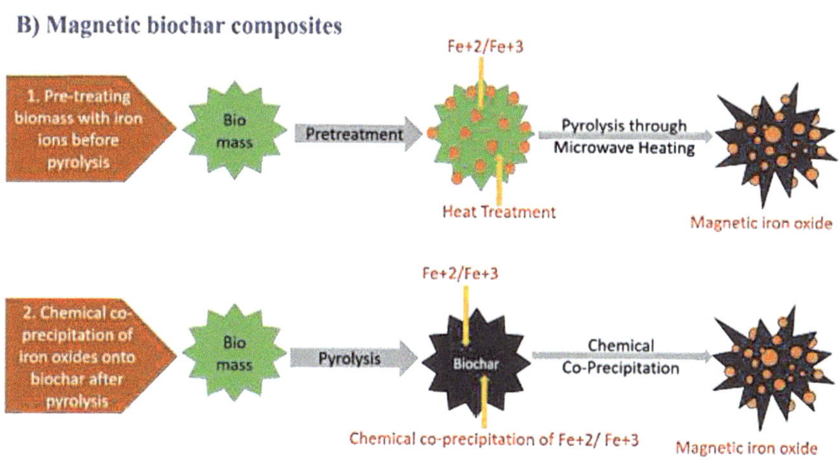

FIGURE 2.2(b) Fabrication methods to construct the magnetic iron oxide nanoparticles.

2.3.3 MULTIFUNCTIONAL NANOPARTICLES-MEDIATED BIOCHAR

Once fully functioning nanomaterials like chitosan, graphene, zinc sulfate, nanoscale zero-valent iron, and carbon nanotubes are coated on the surface of biochar, its outcome is a metal matrix material that would be both cost-effective as well as capable of removing a wide variety of contaminants. This is achieved by incorporating the advantages of the biochar mixture with those of the functional nanoparticle (Figure 2.2(C)). Such beneficial nanomaterials have the potential to considerably boost the surface characteristics, surface area, permeability, and thermal properties of biochar, which would result in an increase in biochar's capacity to remove environmental contaminants [50].

The formation of these functional nanoparticles can be accomplished through either pre-coating multifunctional nanomaterial produced from biomass or the adsorption of functional nanoparticles after combustion. Each of the approaches is described in more detail below. Using a dip-coating method, feedstock could be transformed into biochar-based multifunctional nanocomposites. This conversion takes place fundamentally. Throughout, active nanomaterial mixtures have been made by dispersing bioactive nanostructures in distilled water. The homogenization of resultant mixtures was performed via an ultrasonicator. Following this step, the biomass was kept in the suspension, further dried in an oven, and processed for pyrolysis. After treating biomass feedstock with graphene and pyrene derivatives, a new type of graphene-coated biochar was produced through the process of annealing. Overall findings of experiments demonstrated that graphene significantly enhances the thermodynamic endurance of biochar and also exhibits outstanding methylene blue retention capabilities [59]. Prior to performing pyrolysis, multi-walled carbon nanotube (CNT)-coated biochar was produced through dipping feedstock in mixtures containing varying percentages of carboxyl-functionalized CNTs [60].

The Coating of functional nanoparticles after the pyrolysis method frequently made use of hydrogel, Mg/Al LDH, chitosan, zerovalent iron (ZVI), and ZnS nanocrystals. On the carbon surface of the biochar matrix, these functional nanomaterials were deposited, and they may act as the adsorption positions to remove impurities from solutions. The resulting nanomaterials are more effective due to the presence of both biochar and functional nanoparticle features [50].

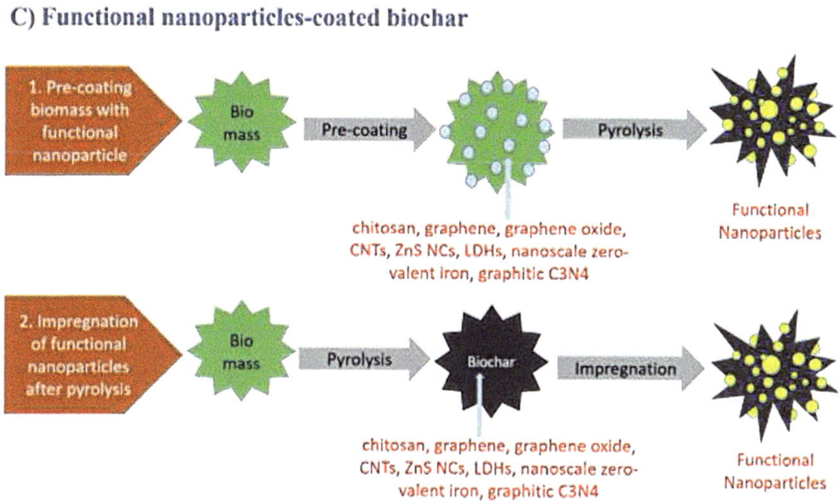

FIGURE 2.2(C) Fabrication methods to construct the functional nanoparticles.

2.4 APPLICATIONS OF NANO-BC (FIGURE 2.3)

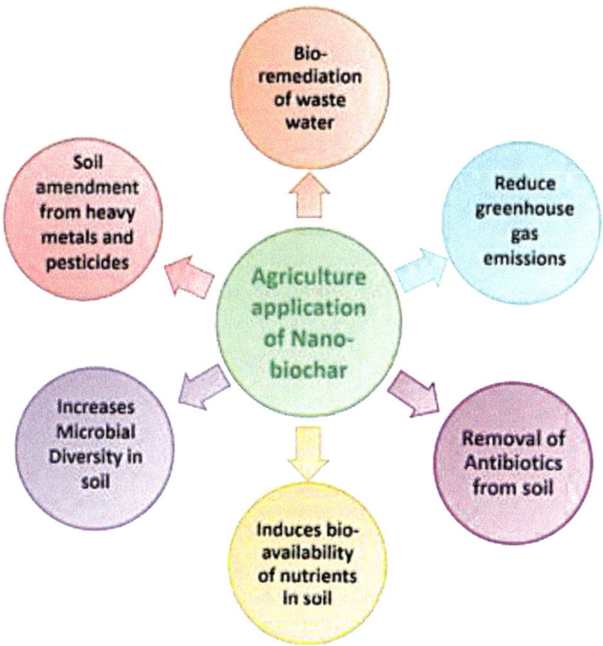

FIGURE 2.3 Applications of nano-biochar to improve soil quality and sustainability.

2.4.1 EFFECT OF NANO-BIOCHAR ON HEAVY METALS (HMS) CONTAMINATED SOILS

As a result of anthropogenic activities, heavy metal (HM) deposition on terrestrial soil has emerged as a severe environmental issue on a global scale. Additionally, HMs can have a negative impact on soil quality and reduce the usefulness of agricultural soils, which will lead to food insecurity [61]. A growing concern is the presence of cadmium (Cd) in soils and its transfer to plants, causing phytotoxicity and health risks to humans through the food chain. According to reports, biochar is a soil amendment that can lower the bioavailability of metals in soil through ionic exchange, electrostatic interactions, and binding of metal ions with ligands present on the surface. The dynamics of Cd in soil were examined in petri dish and pot tests to evaluate the impacts of Cd and nano-biochar on plant growth. Biochar was used in concentrations ranging from 0% to 1%. The concentration of Cd significantly decreased the germination, seed length, and biomass of the four crops (tomato, carrot, lettuce, and cucumber) used in the study. The results showed that germination increased by 11%, 10%, 17.5%, and 10.3%, respectively, when grown on filter paper treated with 500 mg/l nano-biochar. The nano-biochar-treated soils showed a lower concentration of Cd in *Brassica chinensis* L compared to the untreated soils [62].

The effects of metal nanoparticles in reducing the harmful effects of HMs can be exploited by using them in combination with biochar. In a study, the effect of cow-manure biochar, rice straw biochar, foliar ZnO nanoparticles, and a combination of all three, was investigated on sunflowers grown on HM-polluted land. The study focused on the concentration and accumulation of HMs in soil and biomass and the growth, productivity, and quality of the flowers. In comparison to treatments using ZnO NPs alone, the combined treatment greatly reduced the availability of HMs and, as a result, significantly reduced the uptake of HMs by the sunflower. In comparison to untreated plots, the combination treatment decreased the availability of Cd, Cu, Cr, and Pb in the soil by 178.5%, 153.3%, 115.3%, and 78.6%, respectively. Moreover, in plant biomass, it decreased the concentration of Cd, Cu, Cr, and Pb by 0.26, 3.88, 5.19, and 1.13 mg kg_{-1} DM, respectively. The productivity and quality of sunflowers were also significantly improved by the combination treatment. The combined treatment produced results that were 42.6%, 47.0%, and 50.4%

higher than those from the untreated plots in terms of seed production, weight, and number of seeds, respectively [61].

2.4.2 NANO-BC EFFECT ON DIVERSITY OF SOIL MICROBES

The nano-BC interaction with soil may have a significant effect on the biological characteristics of soil that are substantially influenced by a wide variety of microorganisms. Such an interaction may control different processes associated with microorganisms, which would affect how the soil functions [38]. The uptake of macronutrients like phosphorus is amplified by the interactions of nano-BC with minerals present in the soil [63]. Such modifications to the nutrients in the soil may be the main cause of changes in the amount of soil microorganisms [38]. The use of nano-BC in Cd-contaminated soil has produced positive results in terms of microbial richness. With an increasing concentration of nano-BC from 0.0–1%, the diversity of soil microbes has been improved. When compared to control groups, nano-BC enabled an increase in the variety of microorganisms, actinobacteria, and Bacteroides, as well as in their biomass, availability, and species distribution. Nano-BC improves the soil properties alteration like permeability, water retaining capacities, and ductile strength attributed to the modulation in microbial species [62].

2.4.3 NANO-BC-MEDIATED BIOAVAILABILITY OF NUTRIENTS

Soil erosion and loss of nutrients cause depletion of soil fertility. Nano-biochar was applied to enhance crop yield and the quality of the soil. In several experiments, it was seen that various concentrations of the nano-biochar influence several characteristic features of soil like moisture, nutrients, growth of plants, as well as yields of plants. The outcomes demonstrate that soil bulk density was improved along with increased concentration of nano-biochar. Nano-biochar also increases water-holding retention capacity, grain weight, and crop yield, as well as enhances potassium availability in soil [64]. Soil nutrient levels may be negatively impacted by the spread of invasive plant species [65, 66]. Consequently, this prevents the growth of native plants [67, 68].

The spread of the plant *Imperata cylindrica* has put the cultivation of rice in numerous US states in danger. In a study, nano-BC was employed to lessen the stress caused by *I. cylindrica's* root exudates. One of the main chemicals found in *I. cylindrica* root exudates is ferulic acid (FA). FA-treated rice seedlings, when treated with nano-BC, experience detoxification. After applying nano-BC to FA-treated rice seedlings, improvements were seen in their growth. In the pot experiment, the addition of 200 mg kg_{-1} and 400 mg kg_{-1} of nano-BC caused the biomass of the seedlings to rise by 344% and 435%, respectively, in contrast to the FA-treated seedlings. Levels of malondialdehyde (a lipid peroxidation marker) and superoxide dismutase (antioxidant system index) both decrease with the application of nano-BC. In the hydroponic and pot studies, the total chlorophyll content shows a recovery tendency, and the recovery rates are 55% and 96%, respectively, after the addition of 400 mg L_{-1} and 400 mg kg_{-1} of nano-BC to the FA treatments. It can be concluded from this work that soil nutrients needed for the growth of rice seedlings can be restored by nano-BC [69].

2.4.4 ANTIBIOTICS REMOVAL BY NANO-BC

During the past few decades, unbalanced chemical fertilizer applications have significantly worsened the condition of the soil. A sustainable alternative to conventional agriculture is provided by organic methods [70]. Farmers employ manures, such as sewage sludge, and animal, and crop wastes, to improve soil fertility and agricultural output [71]. In some countries, using human urine as fertilizer for agricultural activities has grown in popularity over the last few years [72, 73]. The proportions of nitrogen, phosphorus, and potassium found in human urine are 80%, 50%, and 50%, respectively, of what is required for agriculture [74, 75]. But in addition to nutrients, urine also contains antibiotics, which are widely used by people to treat a range of illnesses. The human body partially digests and absorbs antibiotics, and between 30% and 90% of them are eliminated through urine after 8 to 24 hours of ingestion [76]. Antibiotics must be removed from urine before it can be used as an organic fertilizer. Oxytetracycline has been found to be present in the environment at levels as high as 5–5.7 μg/L due to its discharge through urine [77].

Organic material from bulk biochar must have been utilized to adsorb pharmaceuticals like sulfamethoxazole as well as chloramphenicol [78]. However, nano biochar has an advantage over bulk biochar due to the presence of a higher number of exterior chemical groups and a stronger negative zeta potential [79]. Over the course of five hours, oxytetracycline and other micronutrients extracted from urine were also utilized in the synthesis of nanobiochar [80]. Nanobiochar has more chemical and physical properties due to its high exterior carbon richness and porous structure compared to bulk biochar. Similar results were obtained with disc-milled graphitic nano biochar for the adsorption-based sequestration of oxytetracycline from water, which showed a significant partition coefficient [80]. Another antibiotic, ciprofloxacin, is absorbed by the digestive tract significantly less than other fluoroquinolones [81]. The glomerular filtration rate of ciprofloxacin is higher than usual, which causes a higher amount of renal tubular secretion and a comparatively high ciprofloxacin concentration in urine [82]. Using pomegranate peel-derived nanostructured activated biochar (nPPAB) as a promising sorbent for effectively removing ciprofloxacin has proven to be effective. In batch and packed-bed reactor techniques, the removal efficiency of ciprofloxacin from wastewater by nanostructured activated biochar was 89.94% and 84.74%, respectively (Table 2.1) [83, 88–93, 95–99, 101, 103–105].

2.5 CONCLUSION

Humans depend on the soil for their needs in terms of food, fiber, and feed for cattle, therefore, making soil the fundamental component of life. For many years, anthropogenic activities have highly affected soil health, resulting in soil degradation. Several methods have been practiced to improve soil health, such as mulching, cover crops, minimizing tillage, contour bonds, planting trees, and many more. Nano-biochar is considered a novel approach to sustainable soil health management. In recent years, there has been an expansion in the extent of research conducted into the fabrication of nano-biochar as well as its applicability. The production of nano-biochar is achieved by different techniques, like ball-milling, disc-milling, and pyrolytic methods. Nano-biochar possesses a number of remarkable properties, including highly porous ratios, a substantial majority of surface-active regions, and better thermal and mechanical stability. These are only a few of the outstanding aspects of this material. These characteristics offer wide applications in improving soil remediation, bioavailability of nutrients, and modulating microflora richness,

TABLE 2.1 Different Applications of Nano-Biochar

Sl. No.	Natural Material Used	Nano Product	Type of Method Used	Application	References
1.	Pomegranate peel	Green synthesized nano-BC.	Planetary ball milling	Ciprofloxacin removed from actual wastewater	[83]
2.	Pine-wood	Green synthesized nano-biochar.	Planetary ball milling.	Carbamazepine removal from wastewater.	[45]
3.	*Gliricidia sepium* woody biochar.	Graphitic nano-biochar	Disc-milling	Hexavalent chromium, oxytetracycline, glyphosate, and cadmium removal.	[48]
4.	Rice-husk	Iron oxide permeated mesoporous rice-husk nano-biochars (IPMN).	Pyrolytic method	Arsenic removal	[46]
5.	Wheat straw	Magnetic nano-biochar	Ball-milling	Removal of tetracycline and Hg (II).	[47]
6.	Pinewood biomass and hematite mineral.	Magnetic biochar	Pyrolytic method	Removal of arsenic	[49]
7.	Dip-coating hickory or bagasse biomass.	Carbon nanotube-based nano-biochar.	Pyrolytic method	lead and sulfapyridine absorption and co-absorption.	[84]
8.	Corn stover	ZnS/ZnO-mediated biochar.	Pyrolytic method	Elimination of Cr (VI), Cu (II), and Pb (II).	[85]
9.	Hickory wood	CuO-derived biochar nanocomposites.	Ball milling	Elimination of reactive red 120 (RR120) dye.	[86]
10.	Artichoke leaves	Nano-biochar with surface modified with sodium hydroxide (NaOH).	Pyrolytic method	Elimination of metformin hydrochloride drug.	[87]

hence improving crop growth. However, to recommend nano-biochar for sustainable agricultural practices, further research that focuses on production escalation, understanding of physical and chemical features, addressing potential toxicities, mobilization to different soil regions, and developing further value-added applications is necessary.

ACKNOWLEDGEMENT

The research was supported by the Ministry of Science and Higher Education of the Russian Federation, agreement № 075-15-2023-587.

KEYWORDS

- **agriculture**
- **carbon nanotube**
- **colloidal biochar**
- **municipal solid waste**
- **nano biochar**
- **nanobiochar composites**
- **nano-metal oxide-biochar composites**
- **polychlorinated biphenyls**
- **polycyclic aromatic hydrocarbons**
- **soil organic carbon**
- **soil remediation**
- **specific surface areas**

REFERENCES

1. Shahane, A. A., & Shivay, Y. S. (2021). Soil health and its improvement through novel agronomic and innovative approaches. *Frontiers in Agronomy, 3*, 680456.
2. Gabhane, J. W., Bhange, V. P., Patil, P. D., Bankar, S. T., & Kumar, S. (2020). Recent trends in biochar production methods and its application as a soil health conditioner: A review. *SN Applied Sciences, 2*, 1–21.

3. Nasir Ahmad, N. S. B., Mustafa, F. B., Muhammad Yusoff, S. Y., & Didams, G. (2020). A systematic review of soil erosion control practices on the agricultural land in Asia. *International Soil and Water Conservation Research, 8*, 103–115.
4. Igwe, P. U., Nwezi, C. C., Echendu, J. K., Chukwunyere, I. C., & Okonkwo, N. J. (2017). Adaptations to soil erosion: A review. *International Journal of Advanced Engineering, Management and Science (IJAEMS), 3*(12), 1126–1134.
5. Matuštík, J., Hnátková, T., & Kočí, V. (2020). Life cycle assessment of biochar-to-soil systems: A review. *Journal of Cleaner Production, 259*, 120998.
6. Diatta, A. A., Fike, J. H., Battaglia, M. L., Galbraith, J. M., & Baig, M. B. (2020). Effects of biochar on soil fertility and crop productivity in arid regions: A review. *Arabian Journal of Geosciences, 13*, 1–17.
7. Weber, K., & Quicker, P. (2018). Properties of biochar. *Fuel, 217*, 240–261.
8. Lin, Y., Munroe, P., Joseph, S., Henderson, R., & Ziolkowski, A. (2012). Water extractable organic carbon in untreated and chemical-treated biochars. *Chemosphere, 87*(2), 151–157.
9. Cheng, C. H., Lehmann, J., & Engelhard, M. H. (2008). Natural oxidation of black carbon in soils: Changes in molecular form and surface charge along a climosequence. *Geochimica et Cosmochimica Acta, 72*(6), 1598–1610.
10. Joseph, S. D., Camps-Arbestain, M., Lin, Y., Munroe, P., Chia, C. H., Hook, J., & Amonette, J. E. (2010). An investigation into the reactions of biochar in soil. *Soil Research, 48*(7), 501–515.
11. Bruun, E. W., Hauggaard-Nielsen, H., Ibrahim, N., Egsgaard, H., Ambus, P., Jensen, P. A., & Dam-Johansen, K. (2011). Influence of fast pyrolysis temperature on biochar labile fraction and short-term carbon loss in a loamy soil. *Biomass and Bioenergy, 35*(3), 1182–1189.
12. Ameloot, N., De Neve, S., Jegajeevagan, K., Yildiz, G., Buchan, D., Funkuin, Y. N., & Sleutel, S. (2013). Short-term CO_2 and N_2O emissions and microbial properties of biochar amended sandy loam soils. *Soil Biology and Biochemistry, 57*, 401–410.
13. Sigua, G. C., Novak, J. M., Watts, D. W., Johnson, M. G., & Spokas, K. (2015). Efficacies of designer biochars in improving biomass and nutrient uptake of winter wheat grown in a hard setting subsoil layer. *Chemosphere, 142*, 176–183.
14. Zornoza, R., Moreno-Barriga, F., Acosta, J. A., Muñoz, M. A., & Faz, A. (2016). Stability, nutrient availability, and hydrophobicity of biochars derived from manure, crop residues, and municipal solid waste for their use as soil amendments. *Chemosphere, 144*, 122–130.
15. Schulz, H., & Glaser, B. (2012). Effects of biochar compared to organic and inorganic fertilizers on soil quality and plant growth in a greenhouse experiment. *Journal of Plant Nutrition and Soil Science, 175*, 410–422.
16. Clough, T., Condron, L., Kammann, C., & Müller, C. (2013). A review of biochar and soil nitrogen dynamics. *Agronomy, 3*, 275–293.
17. Ding, Y., Liu, Y., Liu, S., Li, Z., Tan, X., Huang, X., & Zheng, B. (2016). Biochar to improve soil fertility: A review. *Agronomy for Sustainable Development, 36*, 1–18.
18. Atkinson, C. J., Fitzgerald, J. D., & Hipps, N. A. (2010). Potential mechanisms for achieving agricultural benefits from biochar application to temperate soils: A review. *Plant and Soil, 337*, 1–18.

19. Mukherjee, A., & Lal, R. (2013). Biochar impacts soil physical properties and greenhouse gas emissions. *Agronomy, 3*, 313–339.
20. Wang, L., Butterfly, C. R., Wang, Y., Herath, H., Xi, Y. G., & Xiao, X. J. (2014). Effect of crop residue biochar on soil acidity amelioration in strongly acidic tea garden soils. *Soil Use and Management, 30*, 119–128.
21. Ahmad, M., Rajapaksha, A. U., Lim, J. E., Zhang, M., Bolan, N., Mohan, D., & Ok, Y. S. (2014). Biochar as a sorbent for contaminant management in soil and water: A review. *Chemosphere, 99*, 19–23. https://doi.org/10.1016/j.chemosphere.2013.10.071.
22. Paz-Ferreiro, J., Plasencia, P., Gascó, G., & Méndez, A. (2017). Biochar from pyrolysis of deinking paper sludge and its use in the remediation of Zn-polluted soils. *Land Degradation & Development, 28*, 355–360. https://doi.org/10.1002/ldr.2612.
23. Zhu, X., Chen, B., Zhu, L., & Xing, B. (2017). Effects and mechanisms of biochar-microbe interactions in soil improvement and pollution remediation: A review. *Environmental Pollution, 227*, 98–115. https://doi.org/10.1016/j.envpol.2017.04.032.
24. Jeffery, S., Verheijen, F. G. A., van der Velde, M., & Bastos, A. C. (2011). A quantitative review of the effects of biochar application to soils on crop productivity using meta-analysis. *Agriculture, Ecosystems & Environment, 144*, 175–187. https://doi.org/10.1016/j.agee.2011.08.015.
25. Mia, S., van Groenigen, J. W., van de Voorde, T. F. J., Oram, N. J., Bezemer, T. M., Mommer, L., & Jeffery, S. (2014). Biochar application rate affects biological nitrogen fixation in red clover conditional on potassium availability. *Agriculture, Ecosystems & Environment, 191*, 83–91. https://doi.org/10.1016/j.agee.2014.03.011.
26. Ali, S., Rizwan, M., Qayyum, M. F., Ok, Y. S., Ibrahim, M., Riaz, M., & Shahzad, A. N. (2017). Biochar soil amendment on alleviation of drought and salt stress in plants: A critical review. *Environmental Science and Pollution Research, 24*, 12700–12712. https://doi.org/10.1007/s11356-017-8904-x.
27. Drake, J. A., Cavagnaro, T. R., Cunningham, S. C., Jackson, W. R., & Patti, A. F. (2016). Does biochar improve the establishment of tree seedlings in saline-sodic soils? *Land Degradation & Development, 27*, 52–59. https://doi.org/10.1002/ldr.2403.
28. Fahad, S., Hussain, S., Saud, S., Hassan, S., Tanveer, M., Ihsan, M. Z., & Huang, J. (2016). A combined application of biochar and phosphorus alleviates heat-induced adversities on the physiological, agronomical, and quality attributes of rice. *Plant Physiology and Biochemistry, 103*, 191–198. https://doi.org/10.1016/j.plaphy.2016.02.011.
29. De Gryze, S., Cullen, M., Durschinger, L., Lehmann, J., Bluhm, D., Six, J., & Suddick, E. (2010). Evaluation of the opportunities for generating carbon offsets from soil sequestration of biochar. *An Issues Paper Commissioned by the Climate Action Reserve, Final Version*, 1–99. https://doi.org/10.13140/RG.2.1.4071.6967.
30. Gao, X., & Wu, H. (2014). Aerodynamic properties of biochar particles: Effect of grinding and implications. *Environmental Science & Technology Letters, 1*(1), 60–64. https://doi.org/10.1021/ez400072t.
31. Lehmann, J., & Joseph, S. (Eds.). (2015). *Biochar for Environmental Management: Science, Technology, and Implementation.* Routledge. https://doi.org/10.4324/9780203762264.

32. Lonappan, L., Rouissi, T., Das, R. K., Brar, S. K., Ramirez, A. A., Verma, M., & Valero, J. R. (2016). Adsorption of methylene blue on biochar microparticles derived from different waste materials. *Waste Management, 49*, 537–544. https://doi.org/10.1016/j.wasman.2015.12.036.
33. Taheran, M., Naghdi, M., Brar, S. K., Knystautas, E. J., Verma, M., Ramirez, A. A., & Valéro, J. R. (2016). Adsorption study of environmentally relevant concentrations of chlortetracycline on pinewood biochar. *Science of the Total Environment, 571*, 772–777. https://doi.org/10.1016/j.scitotenv.2016.07.066.
34. Naghdi, M., Taheran, M., Pulicharla, R., Rouissi, T., Brar, S. K., Verma, M., & Surampalli, R. Y. (2019). Pine-wood derived nano biochar for removal of carbamazepine from aqueous media: Adsorption behavior and influential parameters. *Arabian Journal of Chemistry, 12*(8), 5292–5301. https://doi.org/10.1016/j.arabjc.2016.11.010.
35. Sulaiman, G. M., Mohammed, W. H., Marzoog, T. R., Al-Amiery, A. A. A., Kadhum, A. A. H., & Mohamad, A. B. (2013). Green synthesis, antimicrobial and cytotoxic effects of silver nanoparticles using *Eucalyptus chapmaniana* leaves extract. *Asian Pacific Journal of Tropical Biomedicine, 3*(1), 58–63. https://doi.org/10.1016/S2221-1691(13)60028-6.
36. AkilaKesavan, G. (2014). Nanotechnology and its applications. *SciTech. J., 1*(06), 12–13. https://doi.org/10.13140/RG.2.2.16324.68489.
37. Naghdi, M., Taheran, M., Brar, S. K., Kermanshahi-Pour, A., Verma, M., & Surampalli, R. Y. (2017). Immobilized laccase on oxygen-functionalized nano biochars through mineral acids treatment for removal of carbamazepine. *Science of The Total Environment, 584*, 393–401. https://doi.org/10.1016/j.scitotenv.2017.01.008.
38. Rajput, V. D., Minkina, T., Ahmed, B., Singh, V. K., Mandzhieva, S., Sushkova, S., & Wang, B. (2022). Nano-biochar: A novel solution for sustainable agriculture and environmental remediation. *Environmental Research, 210*, 112891. https://doi.org/10.1016/j.envres.2022.112891.
39. Noreen, S., & Abd-Elsalam, K. A. (2021). Biochar-based nanocomposites: A sustainable tool in wastewater bioremediation. In *Aqua nanotechnology* (pp. 185–200). Elsevier. https://doi.org/10.1016/B978-0-12-818821-2.00008-5.
40. Oleszczuk, P., Ćwikła-Bundyra, W., Bogusz, A., Skwarek, E., & Ok, Y. S. (2016). Characterization of nanoparticles of biochars from different biomass. *Journal of Analytical and Applied Pyrolysis, 121*, 165–172. https://doi.org/10.1016/j.jaap.2016.07.022.
41. Fan, X., Chang, D. W., Chen, X., Baek, J. B., & Dai, L. (2016). Functionalized graphene nanoplatelets from ball milling for energy applications. *Current Opinion in Chemical Engineering, 11*, 52–58. https://doi.org/10.1016/j.coche.2016.01.004.
42. Chausali, N., Saxena, J., & Prasad, R. (2021). Nano biochar and biochar-based nanocomposites: Advances and applications. *Journal of Agriculture and Food Research, 5*, 100191. https://doi.org/10.1016/j.jafr.2021.100191.
43. Wang, D., Zhang, W., Hao, X., & Zhou, D. (2013). Transport of biochar particles in saturated granular media: Effects of pyrolysis temperature and particle size. *Environmental Science & Technology, 47*(2), 821–828. https://doi.org/10.1021/es303794d.

44. Gámiz, B., Cox, L., Hermosín, M. C., Spokas, K., & Celis, R. (2017). Assessing the effect of organoclays and biochar on the fate of abscisic acid in soil. *Journal of Agricultural and Food Chemistry, 65*(1), 29–38. https://doi.org/10.1021/acs.jafc.6b04426.
45. Charkhi, A., Kazemian, H., & Kazemeini, M. (2010). Optimized experimental design for natural clinoptilolite zeolite ball milling to produce nano powders. *Powder Technology, 203*(2), 389–396. https://doi.org/10.1016/j.powtec.2010.06.006.
46. Shah, U. V., Wang, Z., Olusanmi, D., Narang, A. S., Hussain, M. A., Tobyn, M. J., & Heng, J. Y. (2015). Effect of milling temperatures on surface area, surface energy, and cohesion of pharmaceutical powders. *International Journal of Pharmaceutics, 495*(1), 234–240. https://doi.org/10.1016/j.ijpharm.2015.08.017.
47. Naghdi, M., Taheran, M., Brar, S. K., Rouissi, T., Verma, M., Surampalli, R. Y., & Valero, J. R. (2017). A green method for the production of nano biochar by ball milling optimization and characterization. *Journal of Cleaner Production, 164*, 1394–1405. https://doi.org/10.1016/j.jclepro.2017.07.015.
48. Ramanayaka, S., Tsang, D. C., Hou, D., Ok, Y. S., & Vithanage, M. (2020). Green synthesis of graphitic nano biochar for the removal of emerging contaminants in aqueous media. *Science of the Total Environment, 706*, 135725. https://doi.org/10.1016/j.scitotenv.2019.135725.
49. Nath, B. K., Chaliha, C., & Kalita, E. (2019). Iron oxide permeated mesoporous rice-husk nano biochar (IPMN) mediated removal of dissolved arsenic (As): Chemometric modeling and adsorption dynamics. *Journal of Environmental Management, 246*, 397–409. https://doi.org/10.1016/j.jenvman.2019.05.097.
50. Tan, X. F., Liu, Y. G., Gu, Y. L., Xu, Y., Zeng, G. M., Hu, X. J., & Li, J. (2016). Biochar-based nano-composites for the decontamination of wastewater: A review. *Bioresource Technology, 212*, 318–333. https://doi.org/10.1016/j.biortech.2016.04.093.
51. Yao, Y., Gao, B., Chen, J., & Yang, L. (2013). Engineered biochar reclaiming phosphate from aqueous solutions: Mechanisms and potential application as a slow-release fertilizer. *Environmental Science & Technology, 47*(15), 8700–8708. https://doi.org/10.1021/es4012977.
52. Zhao, C., Wang, B., Theng, B. K., Wu, P., Liu, F., Wang, S., & Zhang, X. (2021). Formation and mechanisms of nano-metal oxide-biochar composites for pollutants removal: A review. *Science of the Total Environment, 767*, 145305. https://doi.org/10.1016/j.scitotenv.2020.145305.
53. Zhang, M., Gao, B., Varnoosfaderani, S., Hebard, A., Yao, Y., & Inyang, M. (2013). Preparation and characterization of a novel magnetic biochar for arsenic removal. *Bioresource Technology, 130*, 457–462. https://doi.org/10.1016/j.biortech.2012.12.104.
54. Mohan, D., Kumar, S., & Srivastava, A. (2014). Fluoride removal from groundwater using magnetic and nonmagnetic corn stover biochars. *Ecological Engineering, 73*, 798–808. https://doi.org/10.1016/j.ecoleng.2014.09.029.
55. Han, Z., Sani, B., Mrozik, W., Obst, M., Beckingham, B., Karapanagioti, H. K., & Werner, D. (2015). Magnetite impregnation effects on the sorbent properties of activated carbons and biochars. *Water Research, 70*, 394–403. https://doi.org/10.1016/j.watres.2014.11.050.

56. Ren, J., Li, N., Li, L., An, J. K., Zhao, L., & Ren, N. Q. (2015). Granulation and ferric oxide loading enable biochar derived from cotton stalks to remove phosphate from water. *Bioresource Technology, 178*, 119–125. https://doi.org/10.1016/j.biortech.2014.08.118.
57. Sun, P., Hui, C., Azim Khan, R., Du, J., Zhang, Q., & Zhao, Y. H. (2015). Efficient removal of crystal violet using Fe_3O_4-coated biochar: The role of the Fe_3O_4 nanoparticles and modeling study of their adsorption behavior. *Scientific Reports, 5*, 1–12. https://doi.org/10.1038/srep12638.
58. Baig, S. A., Zhu, J., Muhammad, N., Sheng, T., & Xu, X. (2014). Effect of synthesis methods on magnetic Kans grass biochar for enhanced As (III, V) adsorption from aqueous solutions. *Biomass and Bioenergy, 71*, 299–310. https://doi.org/10.1016/j.biombioe.2014.10.017.
59. Zhang, M., Gao, B., Yao, Y., Xue, Y., & Inyang, M. (2012). Synthesis, characterization, and environmental implications of graphene-coated biochar. *Science of the Total Environment, 435*, 567–572. https://doi.org/10.1016/j.scitotenv.2012.07.032.
60. Inyang, M., Gao, B., Zimmerman, A., Zhang, M., & Chen, H. (2014). Synthesis, characterization, and dye sorption ability of carbon nanotube–biochar nanocomposites. *Chemical Engineering Journal, 236*, 39–46. https://doi.org/10.1016/j.cej.2013.09.034.
61. Seleiman, M. F., Alotaibi, M. A., Alhammad, B. A., Alharbi, B. M., Refay, Y., & Badawy, S. A. (2020). Effects of ZnO nanoparticles and biochar of rice straw and cow manure on characteristics of contaminated soil and sunflower productivity, oil quality, and heavy metals uptake. *Agronomy, 10*(6), 790. https://doi.org/10.3390/agronomy10060790.
62. Liu, W., Li, Y., Feng, Y., Qiao, J., Zhao, H., Xie, J., & Liang, S. (2020). The effectiveness of nano biochar for reducing phytotoxicity and improving soil remediation in cadmium-contaminated soil. *Scientific Reports, 10*(1), 858. https://doi.org/10.1038/s41598-020-57751-5.
63. Mahmoud, E., El Baroudy, A., Ali, N., & Sleem, M. (2020). Spectroscopic studies on the phosphorus adsorption in salt-affected soils with or without nano-biochar additions. *Environmental Research, 184*, 109277. https://doi.org/10.1016/j.envres.2020.109277.
64. Yang, Y., Zhou, B., Hu, Z., & Lin, H. (2020). The effects of nano-biochar on maize growth in northern Shaanxi province on the loess plateau. *Applied Ecology and Environmental Research, 18*(2), 2863–2877. https://doi.org/10.15666/aeer/1802_28632877.
65. Aerts, R., Ewald, M., Nicolas, M., Piat, J., Skowronek, S., Lenoir, J., & Honnay, O. (2017). Invasion by the alien tree *Prunus serotina* alters ecosystem functions in a temperate deciduous forest. *Frontiers in Plant Science, 8*, 179. https://doi.org/10.3389/fpls.2017.00179.
66. Antunes, C., Pereira, A. J., Fernandes, P., Ramos, M., Ascensão, L., Correia, O., & Máguas, C. (2018). Understanding plant drought resistance in a Mediterranean coastal sand dune ecosystem: Differences between native and exotic invasive species. *Journal of Plant Ecology, 11*(1), 26–38. https://doi.org/10.1093/jpe/rtx007.

67. Zhang, K. M., Shen, Y., Fang, Y. M., Bhowmik, P. C., & Li, Y. (2017). Effects of Bidens Pilosa shoot parts on chlorophyll fluorescence in *Pteris multifida* gametophyte. *Allelopathy Journal, 42*(2), 231–238.
68. Morris, C., Call, C. A., Monaco, T. A., Grossl, P. R., & Dewey, S. A. (2006). Evaluation of elemental allelopathy in *Acroptilon repens* (L.) DC. (Russian Knapweed). *Plant and Soil, 289*, 279–288. https://doi.org/10.1007/s11104-006-9131-5.
69. Shen, Y., Tang, H., Wu, W., Shang, H., Zhang, D., Zhan, X., & Xing, B. (2020). Role of nano-biochar in attenuating the allelopathic effect from *Imperata cylindrica* on rice seedlings. *Environmental Science: Nano, 7*(1), 116–126. https://doi.org/10.1039/C9EN00702B.
70. Karak, T., & Bhattacharyya, P. (2010). Heavy metal accumulation in soil amended with roadside pond sediment and uptake by winter wheat (*Triticum aestivum* L. cv. PBW 343). *The Scientific World Journal, 10*, 2314–2329. https://doi.org/10.1100/tsw.2010.221.
71. Ahadi, N., Sharifi, Z., Hossaini, S. M., Rostami, A., & Renella, G. (2020). Remediation of heavy metals and enhancement of fertilizing potential of sewage sludge by the synergistic interaction of woodlice and earthworms. *Journal of Hazardous Materials, 385*, 121573. https://doi.org/10.1016/j.jhazmat.2019.121573.
72. Rodhe, L., Richert Stintzing, A., & Steineck, S. (2004). Ammonia emissions after application of human urine to a clay soil for barley growth. *Nutrient Cycling in Agroecosystems, 68*(2), 191–198. https://doi.org/10.1023/B:FRES.0000019467.30393.ad.
73. Winker, M., Clemens, J., Reich, M., Gulyas, H., & Otterpohl, R. (2010). Ryegrass uptake of carbamazepine and ibuprofen applied by urine fertilization. *Science of the Total Environment, 408*(8), 1902–1908. https://doi.org/10.1016/j.scitotenv.2010.01.021.
74. Bischel, H. N., Duygan, B. D. Ö., Strande, L., McArdell, C. S., Udert, K. M., & Kohn, T. (2015). Pathogens and pharmaceuticals in source-separated urine in eThekwini, South Africa. *Water Research, 85*, 57–65. https://doi.org/10.1016/j.watres.2015.08.022.
75. Zhou, X., Cuasquer, G. J. P., Li, Z., Mang, H. P., & Lv, Y. (2021). Occurrence of typical antibiotics, representative antibiotic-resistant bacteria, and genes in fresh and stored source-separated human urine. *Environment International, 146*, 106280. https://doi.org/10.1016/j.envint.2020.106280.
76. Nascimento, G. N. L., Rosa, D. L., Nishijo, H., & Aversi-Ferreira, T. A. (2011). Validation of a spectrophotometric method to determine ciprofibrate content in tablets. *Brazilian Journal of Pharmaceutical Sciences, 47*(1), 23–29. https://doi.org/10.1590/S1984-82502011000100003.
77. Kaufmann, A., Butcher, P., Maden, K., & Widmer, M. (2007). Ultra-performance liquid chromatography coupled to time-of-flight mass spectrometry (UPLC–TOF): A novel tool for multiresidue screening of veterinary drugs in urine. *Analytica Chimica Acta, 586*(1–2), 13–21. https://doi.org/10.1016/j.aca.2006.12.024.
78. Yang, F., Zhang, Q., Jian, H., Wang, C., Xing, B., Sun, H., & Hao, Y. (2020). Effect of biochar-derived dissolved organic matter on adsorption of sulfamethoxazole and chloramphenicol. *Journal of Hazardous Materials, 396*, 122598. https://doi.org/10.1016/j.jhazmat.2020.122598.

79. Lian, F., Yu, W., Wang, Z., & Xing, B. (2018). New insights into black carbon nanoparticle-induced dispersibility of goethite colloids and configuration-dependent sorption for phenanthrene. *Environmental Science & Technology, 53*(2), 661–670. https://doi.org/10.1021/acs.est.8b04331.
80. Ramanayaka, S., Kumar, M., Etampawala, T., & Vithanage, M. (2020). Macro, colloidal, and nano biochar for oxytetracycline removal in synthetic hydrolyzed human urine. *Environmental Pollution, 267*, 115683. https://doi.org/10.1016/j.envpol.2020.115683.
81. Brumfitt, W. I. D. J. A., Franklin, I. T. A. L. I. A., Grady, D. E. I. D. R. E., Hamilton-Miller, J. M., & Iliffe, A. (1984). Changes in the pharmacokinetics of ciprofloxacin and fecal flora during administration of a 7-day course to human volunteers. *Antimicrobial Agents and Chemotherapy, 26*(5), 757–761. https://doi.org/10.1128/AAC.26.5.757.
82. Aronoff, G. E., Kenner, C. H., Sloan, R. S., & Pottratz, S. T. (1984). Multiple-dose ciprofloxacin kinetics in normal subjects. *Clinical Pharmacology & Therapeutics, 36*(3), 384–388. https://doi.org/10.1038/clpt.1984.204.
83. Hamadeen, H. M., & Elkhatib, E. A. (2022). New nanostructured activated biochar for effective removal of antibiotic ciprofloxacin from wastewater: Adsorption dynamics and mechanisms. *Environmental Research, 210*, 112929. https://doi.org/10.1016/j.envres.2022.112929.
84. Inyang, M., Gao, B., Zimmerman, A., Zhou, Y., & Cao, X. (2015). Sorption and cosorption of lead and sulfapyridine on carbon nanotube-modified biochars. *Environmental Science and Pollution Research, 22*, 1868–1876. https://doi.org/10.1007/s11356-014-3414-1.
85. Li, C., Zhang, L., Gao, Y., & Li, A. (2018). Facile synthesis of nano ZnO/ZnS modified biochar by directly pyrolyzing zinc-contaminated corn stover for Pb (II), Cu (II), and Cr (VI) removals. *Waste Management, 79*, 625–637. https://doi.org/10.1016/j.wasman.2018.08.003.
86. Wei, X., Wang, X., Gao, B., Zou, W., & Dong, L. (2020). Facile ball-milling synthesis of CuO/biochar nanocomposites for efficient removal of reactive red 120. *ACS Omega, 5*(11), 5748–5755. https://doi.org/10.1021/acsomega.9b04063.
87. Mahmoud, M. E., El-Ghanam, A. M., Saad, S. R., & Mohamed, R. H. A. (2020). Promoted removal of metformin hydrochloride anti-diabetic drug from water by fabricated and modified nano biochar from artichoke leaves. *Sustainable Chemistry and Pharmacy, 18*, 100336.
88. Rajput, V. D., Minkina, T., Kumari, A., Harish, Singh, V. K., Verma, K. K., & Keswani, C. (2021). Coping with the challenges of abiotic stress in plants: New dimensions in the field application of nanoparticles. *Plants, 10*(6), 1221.
89. Rajput, V., Minkina, T., Sushkova, S., Behal, A., Maksimov, A., Blicharska, E., & Barsova, N. (2020). ZnO and CuO nanoparticles: A threat to soil organisms, plants, and human health. *Environmental Geochemistry and Health, 42*, 147–158.
90. Rajput, V. D., Minkina, T., Sushkova, S., Tsitsuashvili, V., Mandzhieva, S., Gorovtsov, A., & Gromakova, N. (2018). Effect of nanoparticles on crops and soil microbial communities. *Journal of Soils and Sediments, 18*, 2179–2187.

91. Rajput, V., Minkina, T., Fedorenko, A., Sushkova, S., Mandzhieva, S., Lysenko, V., & Ghazaryan, K. (2018). Toxicity of copper oxide nanoparticles on spring barley (*Hordeum sativum* distichum). *Science of the Total Environment, 645*, 1103–1113.
92. Rajput, V. D., Kumari, A., Minkina, T., Barakhov, A., Singh, S., Mandzhieva, S. S., & Garg, M. C. (2022). A practical evaluation of the integrated role of biochar and nanomaterials in soil remediation processes. *Environmental Geochemistry and Health, 45,* 9435–9449.
93. Rajput, V. D., Faizan, M., Upadhyay, S. K., Kumari, A., Ranjan, A., Sushkova, S., & Deryabkina, I. (2022). Influence of nanoparticles on the plant rhizosphere microbiome. In A. Singh (Ed.), *The Role of Nanoparticles in Plant Nutrition under Soil Pollution: Nanoscience in Nutrient Use Efficiency* (pp. 83–102). Cham: Springer International Publishing.
94. Rajput, V. D., Singh, A., Minkina, T., Rawat, S., Mandzhieva, S., Sushkova, S., & Upadhyay, S. K. (2021). Nano-enabled products: Challenges and opportunities for sustainable agriculture. *Plants, 10*(12), 2727.
95. Singh, A., Singh, A. K., Rawat, S., Pal, N., Rajput, V. D., Minkina, T., & Tripathi, J. N. (2022). Satellite-based quantification of methane emissions from wetlands and rice paddies ecosystems in North and Northeast India. *Hydrobiology, 1*(3), 317–330.
96. Singh, A., Sengar, R. S., Rajput, V. D., Minkina, T., & Singh, R. K. (2022). Zinc oxide nanoparticles improve salt tolerance in rice seedlings by improving physiological and biochemical indices. *Agriculture, 12*(7), 1014.
97. Kanaoujiya, R., Saroj, S. K., Rajput, V. D., Alimuddin, Srivastava, S., Minkina, T., ... & Kumar, A. (2023). Emerging applications of nanotechnology for mankind. *Emergent Materials*, *6*(2), 439–452.
98. Singh, A., Mehrotra, R., Rajput, V. D., Dmitriev, P., Singh, A. K., Kumar, P., & Singh, A. K. (2022). Geoinformatics, artificial intelligence, sensor technology, big data: Emerging modern tools for sustainable agriculture. In A. Singh (Ed.), *Sustainable Agriculture Systems and Technologies* (pp. 295–313). Cham: Springer International Publishing.
99. Singh, A., Rajput, V. D., Rawat, S., Sharma, R., Singh, A. K., Singh, A. K., & Tomar, R. S. (2022). Geoinformatics and nanotechnological approaches for coping up abiotic and biotic stress in crop plants. In A. Singh (Ed.), *Sustainable Agriculture Systems and Technologies* (pp. 337–359). Cham: Springer International Publishing.
100. Singh, A., Rajput, V. D., Rawat, S., Sharma, R., Singh, A. K., Singh, A. K., & Tomar, R. S. (2022). Emerging tools for sustainable agriculture and food security: An introduction. In A. Singh (Ed.), *Sustainable Agriculture Systems and Technologies* (pp. 1–15). Cham: Springer International Publishing.
101. Rajput, P., Rajput, V. D., Singh, A., Sharma, R., Rawat, S., Tomar, R. S., & Minkina, T. (2022). Impact of emerging tools for solving 'hidden hunger' in humans, In: *Emerging Tools for Sustainable Agriculture and Food Security*. Publisher: Deepika Book Agency Publishers & Distributors, India.
102. Rajput, V., Minkina, T., Kumari, A., Sudhir S., S., Ranjan, A., Faizan, M., Barakvov, A., Gromovik, A., Gorbunova, N., Rajput, P., Singh, A., Khabirov, I., Nazarenko, O., Sushkova, S., & Kızılkaya, R., (2022). A review on nanobioremediation approaches for restoration of contaminated soil. *Eurasian J. Soil Sci. 11*(1), 43–60. DOI : 10.18393/ejss.990605.

103. Rajput, V. D., Minkina, T., Feizi, M., Kumari, A., Khan, M., Mandzhieva, S., & Choudhary, R. (2021). Effects of silicon and silicon-based nanoparticles on rhizosphere microbiome, plant stress and growth. *Biology, 10*(8), 791.
104. Singh, A., Rajput, V. D., Rawat, S., Singh, A. K., Bind, A., Singh, A. K., & Lobzenko, I. P. (2020). Monitoring soil salinity and recent advances in the mechanism of salinity tolerance in plants. *Biogeosystem Technique, (7)*, 66–86.
105. Singh, A., Sengar, R. S., Shahi, U. P., Rajput, V. D., Minkina, T., & Ghazaryan, K. A. (2023). Prominent effects of zinc oxide nanoparticles on roots of rice (*Oryza sativa* L.) grown under salinity stress. *Stresses, 3*(1), 33–46.

CHAPTER 3

Nanoparticle, Types, and Approaches for Improving Soil Health

RICHA DAS,[1] PRADEEP KUMAR,[2] SHRENI AGRAWAL,[1] AMIT KUMAR SINGH,[3] PRAVEEN KUMAR SHUKLA,[2] POOJA VERMA,[2] VISHNU D. RAJPUT,[4] INDRANI BHATTACHARYA,[1] SUNIL KUMAR MISHRA,[3] and KAVINDRA NATH TIWARI[2]

[1]*Department of Biotechnology, Parul Institute of Applied Science, Parul University, Vadodara, Gujarat, India*

[2]*Department of Botany, MMV, Banaras Hindu University, Varanasi, Uttar Pradesh, India*

[3]*Department of Pharmaceutical Engineering and Technology, Indian Institute of Technology, Banaras Hindu University, Varanasi, Uttar Pradesh, India*

[4]*Academy of Biology and Biotechnology, Southern Federal University, Rostov on Don, Russia*

ABSTRACT

Soil is Earth's most important natural resource for sustaining life. Soil health is a critical factor in crop development. Nevertheless, as a result of both natural and anthropogenic factors, the soil's health to support crop development and production deteriorates with time. Factors such as excess salinity, drought, unavailability of nutrients due to volatilization or low solubility, and heavy metals impact soil health. These factors also create an imbalance of soil microflora. Irrational use of fertilizers and climatic

Nano-Biochar and Nanoparticles for Soil Health Improvement.
Vishnu D. Rajput and Tatiana M. Minkina (Eds.)

conditions hugely impact soil, leaving a large land unfit for cultivation. Nanotechnological approaches are the prime techniques for soil health improvement. Nanoparticles have several impressive characteristics, like large surface area, nanoporosity, biocompatibility, easy surface functionalization, mechanical and thermal stability, and magnetic, chemical, and optical properties, therefore making them a potential approach for soil health improvement. Nanoparticles are categorized into different types, like inorganic, carbon, organic, and composite-based nanoparticles. There are numerous ways for synthesizing nanoparticles, which are classified as top-down and bottom-up approaches. The main aspect of this review is discussing different nanotechnological approaches to improve soil health. Nano-biochar, a novel organic nanotechnological approach to soil health improvement, has also received attention in sustainable agriculture and is discussed in this review. Furthermore, different methods involved in the fabrication of nanoparticles are addressed, followed by different kinds. The review is then concluded by a few major challenges associated with nanoparticles.

3.1 INTRODUCTION

Soil can generally be defined as a "fertile substrate." Under optimum conditions, minerals, organic matter, water, and oxygen are all in equilibrium in agricultural soils. Each of these elements contributes in a balanced way, allowing for water storage and drainage, oxygen, and nutrient availability to stimulate crop development and the structural stability of plants [1]. Nowadays, soils have become prone to different abiotic stresses, such as strong wind, flooding, and high temperatures. Along with soil erosion, extensive soil usage has led to a global water crisis, droughts, and deforestation, each of which poses a danger to the viability of our agricultural systems [1]. Multiple internal and external effects of soil erosion involve a decrease in the production of agricultural fields, cracking and thinning of the soil surface, poor water holding capacity and nutrient retention, silting in surface water sources that cause flooding, reduction in dry-water flows, and deterioration of environmental sustainability [2]. A crucial tenet required for agriculture will be lost [3] if the misuse of soil nutrients keeps reducing the productivity of the soil and the number of fertile agricultural land [4]. It is clear that activities to stop environmental deterioration have to become a major focus of our global community in order to preserve and grow food production [1].

In order to produce food in a way that can be managed with the least possible harm to biodiversity and environmental resources, a farming method known as sustainable agriculture can be used. This sustainable method of farming aims to preserve soil and other environmental resources while generating commercial success and upholding social equality [5]. Some of the conventional approaches include contoured planting, mulching with biomass waste, and structural constructions like check dams can be utilized to enhance soil quality. To prevent soil erosion, advanced techniques such as soil netting, the spreading of stone blankets, the use of biodegradable polymers, geosynthetic substances, mulching, chemical soil binders, changed weeding techniques in agricultural fields, and other enhanced agronomic practices can be utilized [2].

Nanotechnology is a developing science with the potential to be utilized in a wide range of disciplines. The restoration of the damaged soil quality may be possible using nano-enabled soil remediation. Nanotechnology applications are economical, simple to implement, and offer productive treatment processes that might drastically reduce soil pollution [6]. The interest in employing nanoparticles (NPs) for restoring environmental health is based on the qualities of the nanoparticle. Nanoparticles (NPs) exhibit catalytic action, excellent electrical characteristics, large surface-area-to-mass ratios, and great sensitivity [7]. Furthermore, the large surface area, random distribution of catalytic sites, and numerous covering modification options allow the use of NPs in adsorption-based removal processes [8]. Additionally, NPs can spread in restricted spaces, enhancing their application in soil and water cleansing. Hyperaccumulators and native soil bacteria may be able to speed up the biodegradation process with the help of NPs, potentially expanding the scope of recovery. This is also known as microbial-mediated nano-remediation and nano-phytoremediation [9].

NPs have been created using a variety of ways, including chemical and physical processes. The aim is particularly to choose techniques for fabricating NPs that are non-toxic and ecologically friendly. A collection of biological entities like fungi, plants, actinomycetes, algae, bacteria, and yeast has emerged as the greatest contender for fabricating NPs biologically. This fabricating technique is often called "green synthesis" [10].

Top-down and bottom-up are the two approaches of fabrication. The top-down method refers to the process of breaking down large materials into small particles. It primarily consists of a big tube furnace used for crushing the bulk matter into much smaller particles. Thermal decomposition, grinding, ball milling, sputtering, mechanical milling, laser ablation, and lithography are the primary top-down processes. While the bottom-up approach proposes that NPs be produced from smaller molecules to the needed substance. In this process, chemical reduction is typically used, occasionally in combination with a capping agent for the stabilization of produced NPs. Spinning, pyrolysis, chemical vapor deposition, and the sol-gel method are the frequent bottom-up techniques employed. The biological technique of fabrication, in which biomolecules mix with metallic elements to generate materials at the nanoscale, is also categorized as a bottom-up approach [11]. NPs can be generally classified into a number of groups based on their size, morphology, and chemical properties. They are classified into four types based on their physical and chemical features. These are inorganic, organic, carbon, and nanocomposite-based nanoparticles [12].

Numerous environmental restoration applications can be successfully used with these various types of nanomaterials. To mitigate a specific pollutant, it is important to look for a suitable NP, and this relies on the type of targeted pollutant, the concentration of NP required for successful remediation, and the degree of accessibility of the site that is to be remediated. Although the recyclability of some materials has been discussed, it seems that eventually their effectiveness decreases and they become useless. In order to prevent the risk that these materials would end up becoming a source of environmental pollution, the study is required to clarify what happens to them once they are introduced to the environment for remedial reasons [8]. This review article involves the synthesis of nanomaterials, their types, applications, prospects, and challenges related to nanotechnology for soil improvement and soil remediation.

3.2 NANOPARTICLE FABRICATION

There are several ways for fabricating nanoparticles, which are divided into two categories, top-down and bottom-up techniques.

3.2.1 TOP-DOWN TECHNIQUE

Top-down synthesis of nanomaterials involves the breakdown of bigger molecules (bulk material) into smaller molecules, which subsequently undergo conversion into nanoparticles. Thermal decomposition, grinding or ball milling, mechanical milling, lithography, laser ablation, and sputtering are examples of top-down synthesis methods [13]. This technique was mostly utilized to generate colloidal carbon particles that were spherical and ranged in size from 20 to 50 nm [14].

3.2.1.1 THERMAL DECOMPOSITION METHOD

Also known as thermolysis, thermal decomposition is an endothermic process. It comprises the chemical degradation of substances caused by heat. Heat is essential in this process to disrupt the chemical bonds present in the molecule that is being degraded [15]. At the decomposition temperature, an element gets chemically decomposed [14]. The main purpose of this technique is to generate metal nanoparticles. The pressure and temperature at which thermal decomposition occurs depend on factors like the composition of the metallic ion and the strength of interaction with ligands in conjugated compounds [16]. Ahab et al. employed the thermal decomposition procedure to generate gadolinium oxide nanoparticles and characterized them with paramagnetic polyethylene glycol [17].

3.2.1.2 MECHANICAL-MILLING METHOD/BALL-MILLING METHOD

To generate compounds at the nanoscale from bulk materials, mechanical milling is an effective and economical process [18]. Nanomaterials, nano-composite materials, and nano-quasicrystalline materials might all be synthesized using this process with regulated parameters. Mechanical milling goals include reduction in particle size, particle mixing or blending in new phases, particle shape modifications, and nanocomposite synthesis. The numerous types of ball milling, in which the balls strike upon the powder charge, can be employed for the synthesis of nanomaterials [19]. The energy delivered to the powder from the balls during milling determines the kinetics of mechanical milling or alloying [20]. Many parameters control energy transmission, including mill type, powder delivered to operate the milling chamber, milling velocity, ball size, and size distribution, dry or wet milling, milling temperature, and milling duration [21]. Aluminum alloys that have been reinforced by oxide and carbide, spray coatings that are resistant to wear, nanoalloys based on aluminum, nickel, magnesium, and copper, and a variety of other nanocomposite materials can all be generated by using the mechanical milling method [22]. This method can mainly be used to produce nanocomposite-based nanobiochar from bulk biochar. By using a high-energy planetary ball milling technique, Nandi et al. [23] developed amorphous and/or nanocrystalline Al-based alloys from elemental powder. By using mechanical alloying, Khakbiz et al. [24] constructed the Al-B4C nanocomposite powders.

3.2.1.3 LITHOGRAPHY

Lithography is an effective technique for constructing nanoarchitectures with an intense electron or light beam. Although it is most capable of producing micron-sized features, it is energy-intensive and expensive to manufacture. For more than a decade, this method has been employed to construct printed circuit boards and computers. Numerous lithography methods exist, such as photo-, electron beam, soft, focused ion, nanosphere, and dip pin lithography [14]. A unique low-cost fabrication method called nanosphere lithography (NSL) can be used to generate uniform, consistent arrays of nanoparticles of various sizes. Because colloidal spheres self-organize in an HCP lattice and because structured layers are obtained using a traditional lithography technique, NSL is sometimes viewed as a hybrid between the bottom-up and top-down approaches [25]. Lithography is mainly classified into two types: masked and maskless lithography. In masked type, a specific mask or template is utilized to transfer nanopatterns over a large surface area. In maskless type, random nanopattern printing is performed without the use of a mask [26].

3.2.1.4 LASER ABLATION TECHNIQUE

It is a technique frequently employed for the fabrication of different NPs, such as CNTs, semiconductor-based quantum dots, and nanowires. In this approach, species that have been laser-vaporized proliferate and form nanoparticles in a background gas. This process is primarily utilized to generate highly pure nanoparticles of size < 10 nm [27]. The process of ablating targeted metals begins with the absorption of energy from a laser beam. When the laser beam interacts with the target, heat is generated and the target is photoionized. Following that, metal nanoparticles will be ejected from the metal plate, depending on the absorbed energy [28]. Without the use of chemicals or surfactants, it is an environmentally friendly and straightforward technique for producing metal nanoparticles. The flexibility, purity, ability to synthesize a variety of ceramics and metals, and distribution of NPs in different liquids in situ are benefits of the laser ablation approach [29].

3.2.1.5 SPUTTERING METHOD

It is a method of fabricating nanoparticles by striking solid surfaces with projectiles such as gas or plasma, possessing high energy. This approach is expected to be effective in producing thin films of nanomaterial [30]. It may be accomplished in a number of methods, including the use of radio-frequency diodes, DC diodes, and magnetrons [31]. It is frequently performed in an evacuated chamber that is subsequently filled with sputtering gas. Free valence electrons strike the cathode target at a high voltage, producing gas ions as a result of the collision. The cathode target is continuously struck by the strongly accelerating positively charged ions in the electric field, which causes the ejection of atoms from the target surface [32]. This process produces nanoparticles of different shapes and sizes, which depend on the temperature, layer thickness, annealing time, substrate, etc. [33]. It is a fascinating process because, in contrast to electron-beam lithography, it is more affordable and produces nanomaterials whose composition is similar to the target material but with fewer imperfections [34].

3.2.2 BOTTOM-UP TECHNIQUE

The bottom-up method involves starting with a single atom or molecule and building up to generate nanomaterials. In this technique, the assembly of atoms or molecules into desirable nanostructures results in the controlled segregation of the constituent parts (2–10 nm size range). The chemical vapor deposition process, the sol-gel method, spinning, pyrolysis, and biological synthesis are a few of the techniques employed in the bottom-up approach [35].

3.2.2.1 CHEMICAL VAPOR DEPOSITION (CVD) TECHNIQUE

CVD is a technique often used in semiconductor-based industries to deposit various materials in thin-layer form. The substrate must be exposed to one or more volatile precursors during the procedure [35]. Precursors are considered suitable for CVD if they have adequate volatility, excellent chemical purity, high stability during evaporation, affordability, non-hazardous, and longer lifespan. Moreover, a good precursor, upon decomposition, should not leave any toxins behind [36]. During the process of generating CNTs, for example, a substrate is subjected to high temperatures. As a precursor, hydrocarbons, a carbon-containing gas, are progressively fed into the system. Carbon molecules are generated during the decomposition of a gas when subjected to extreme temperatures, and when they recombine, they form CNTs [37]. However, the catalyst chosen has a considerable impact on the form and kind of nanomaterial generated. During the CVD-based production of the material, Co and Ni catalysts yield multilayer graphene, but Cu, when used as a catalyst, only forms monolayer graphene [38].

3.2.2.2 SOL–GEL TECHNIQUE

Sol-gel is a wet-chemical technique that is routinely used for the fabrication of nanoparticles. This method was designed to produce a diverse range of superior-quality metal oxide nanomaterials. This technique is known as a sol-gel method because the liquid precursor is transformed into a sol during the production of the metal-oxide nanoparticles, and the sol is then modified into a network structure known as a gel [39]. The technique is named sol-gel because, during the synthesis of metal-oxide NP, the transformation of a precursor which is in liquid form is converted into sol. A gel is a semi-solid, rigid, interconnected network of structures in the nano range occurring in a continuous liquid phase. This is in contrast to sol, as it is in liquid form into which solid colloidal particles are distributed evenly. The fabrication of sol-gel films typically involves one of three methods: (i) transformation of colloidal soil particles to gel form; (ii) metal oxide hydrolysis and polycondensation, followed by gel drying; or (iii) metal oxide hydrolysis and polycondensation, followed by aging and dehydrating. This technique can be used to construct nano-films and nano-powders [40]. Precursor type, aging duration, hydrolysis rate, pH, and precursor to water molarity, all have an influence on the resultant material created by the sol-gel technique [41]. It is a low-cost process with various advantages, like the homogeneous nature of the material synthesized, operating under low temperatures, and straightforward procedure [42].

3.2.2.3 PYROLYSIS

Pyrolysis is employed frequently for processing organic wastes thermally to fabricate carbon NPs. Depending on the increasing size, media, and physicochemical characteristics, the method of nanoparticle formation dictates the active distribution of nanoparticles. Various pyrolysis methods are described, with laser pyrolysis being a popular bottom-up method for the production of carbon-based NPs, like carbon nanotubes (CNTs), graphene, and carbon nanofibers (CNFs) [43]. In this method, circulating reactant gases are heated using a constant wave CO_2 laser, which causes chemical breakdown to generate vapors that start the nucleation process and thereafter lead to the development of nanoparticles. When temperatures are high, coalescence is more intense, producing spherical particles, and when temperatures are low, other shapes are produced. Thus, the features of the nanoparticles can be adjusted by changing the operational parameters. It is preferable to use this procedure to create high-quality, small particles with a limited size distribution. Additionally, the method generates high-purity nanomaterial continuously with the least side effects [40].

3.2.3 BIOLOGICAL FABRICATION OF NANOPARTICLES

Biological synthesis is the formation of nanoparticles utilizing plant extracts and microorganisms such as fungi and bacteria. This technique is mostly used to generate metal nanoparticles such as zirconium oxide, titanium, gold, silver, and copper NPs. It has been demonstrated that microorganisms are significant nano factories with enormous potential as environmentally acceptable and economically advantageous tools, avoiding caustic, poisonous substances and the high energy consumption necessary for physiochemical synthesis. Due to a variety of catalytic enzymes that may convert metal salts into metal nanoparticles with a specific size distribution, microorganisms have the capacity to collect and remove heavy metals. Over the past few years, extra- and intracellular studies of microorganisms, such as bacteria (such as actinomycetes), fungi, and yeasts, for the manufacture of metal nanoparticles have been conducted. Bacterial feedstock, supernatant, and derivative materials have been used in a variety of biological procedures for nanoparticle formation.

Phyto nanotechnology has opened up new possibilities for the production of nanoparticles from plant extracts. The elevated requirement for nanoparticles with applications in the biomedical and environmental fields can be satisfied by using plant-derived nanoparticles made from readily available plant materials like leaves, roots, shoots, seeds, and fruits. Plants are also non-toxic, which makes them ideal for this purpose [44]. Proteins, amino acids, organic acids, vitamins, and secondary metabolites like flavonoids, alkaloids, polyphenols, terpenoids, heterocyclic compounds, and polysaccharides have all been suggested to play important roles in the reduction of metal salts and additionally function as encapsulation and stabilizing agents for synthesized nanoparticles [45]. By changing the environment in which they grow or the functional molecules they interact with, this technique can be used to modify the shape and size of metal nanoparticles [44, 46]. The size, shape, and composition of nanoparticles are influenced by a number of variables, including temperature, pH, the concentration of developing microorganisms, and the concentration of plant extract.

Numerous benefits of the biological synthesis of nanoparticles include quick and environmentally friendly production processes, as well as the economical and biodegradable nature of synthesized nanoparticles. Additionally, because plant and microbial elements already serve as encapsulating and stabilizing agents, additional stabilizing agents are not needed [47].

3.3 TYPES OF NANOPARTICLES

3.3.1 INORGANIC NANOPARTICLES

Metal precursors are used to synthesize inorganic nanoparticles. They can be mono-, bi-, or poly-metallic in nature [48]. As its name suggests, monometallic nanoparticles (MNPs) are made up of just one type of metal. The characteristics of these nanoparticles depend on the metal atom that makes them up (Figure 3.1). Monometallic nanoparticles can be metallic, magnetic, and transition metal types [49]. Bimetallic nanoparticles are made up of two different types of metals [48]. Bimetallic nanoparticles are synthesized synchronously causing the reduction of two ions of a metal under suitable stabilizing approaches, such as steric hindrance and electrostatic repulsive forces. The catalytic features of the nanoparticles can be greatly enhanced through bi-metallization, which may not be possible with monometallic nanoparticles. Platinum, nickel, and gold-based bimetallic nanoparticles are some examples [49]. Polymetallic nanoparticles, on the other hand, are those that consist of more than two different types of metals. These nanoparticles have a much larger surface area; therefore, they facilitate a much higher adsorption rate [165]. The size of inorganic nanoparticles is between 1 and 100 nm [50]. They have unique electrical and optical properties because of the localized surface plasmon resonance [12]. Moreover, they also possess magnetic, thermal, and biological properties [51]. The presence of these properties makes them important for nanodevice development, which has wide applications in biological, chemical, physical, pharmaceutical, and biomedical areas [52, 53]. Zinc, aluminum, silver, cadmium, lead, cobalt, iron, gold, and copper are the most often employed metals for nanoparticle synthesis [54]. Metal oxide-based nanoparticles have garnered a lot of attention in recent years due to their unique features. At room temperature and under environmental oxygen, iron nanoparticles oxidize quickly to iron oxide. When compared to metal nanoparticles, this enhances the reactivity of oxide nanoparticles. They possess increased reactivity and efficiency. Zinc oxide, aluminum oxide, titanium oxide, cerium oxide, magnetite, silicon dioxide, and iron oxide are the most frequently synthesized metal oxide nanoparticles [51, 54]. For the fabrication of inorganic nanoparticles, different methods like spray pyrolysis, chemical vapor deposition, sol-gel technique, and quick solidification are commonly used [55].

3.3.2 ORGANIC NANOPARTICLES

They are composed solely of organic substances such as lipids, polymers, carbohydrates, and proteins. Their sizes are between 10 nm and 1 μm

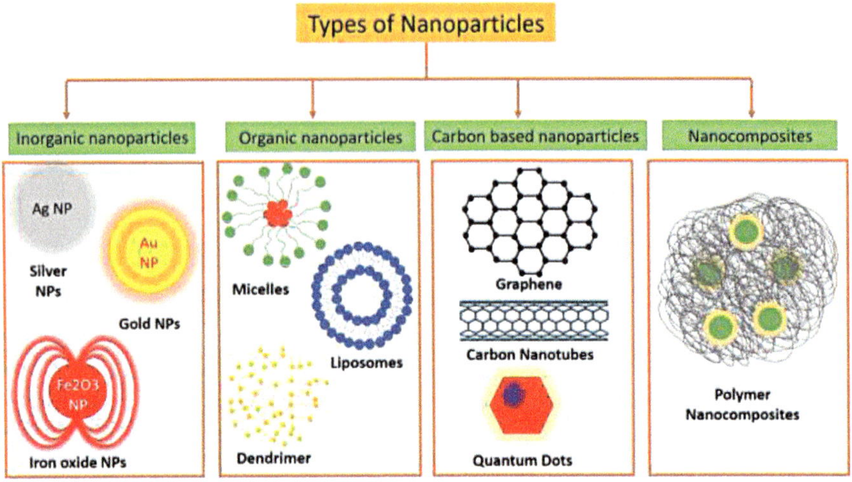

FIGURE 3.1 Diagram representing different types of nanoparticles.

[56, 57]. The unique size effects of organic nanoparticles, like optical, chemical, and physical properties, depend on the structural confinement, like packing, conformation, and elastic properties of organic nanoparticles. The same has been confirmed through spectroscopic data of different organic compounds [58]. The weak interactions holding the nanoparticles together give them a dynamic character. Vesicles and micelles can combine to form larger particles. The continual supply of surfactant to a micellar system can lead the micelles to grow bigger and even change form to cylindrical [59]. They are biodegradable and non-toxic [60]. But, sensitive to electromagnetic and thermal conditions [51]. For encapsulation, these nanoparticles are highly preferred. The mechanical milling method is the popular top-down approach, but more recently, microfluidics and lithography have also been employed to create organic nanomaterials. When using the bottom-up approach, organic NPs are often created through condensation and precipitation techniques that employ synthetic techniques or self-assembly principles [59]. Micelles, chitosan, liposomes, dendrimers, and cellulose are examples, as shown in Figure 3.1 [60].

3.3.3 CARBON-BASED NANOPARTICLES

Carbon-based nanoparticles are formed solely of carbon atoms [51]. Since carbon-based nanoparticles (CBN) have unique properties, such as fluorescence, an adjustable narrow emission spectrum, and good photostability, they are widely used in imaging biological samples. Moreover, their surfaces can be altered with functional groups, such as epoxy, carboxylic acid, and hydroxyl, to improve their properties. CBN is one of the most preferred alternatives for theranostic applications due to its large surface area, and superior optical, and mechanical properties [61]. Graphene oxide, carbon nanotubes, fullerenes, carbon-based quantum dots, and nanodiamonds are some of the types of CBN (Figure 3.1) [62].

CNTs have a diameter of 1 nm, manufactured from a graphite sheet. They are one atom thick, with different characteristics based on the rolling technique [63, 64]. CNTs are exceptionally hard because of their enormous tensile strength [65]. Furthermore, because of their 1D, tubular structure in the nano range, their thermal and electrical conductance improves dramatically and surpasses other conductive metals [66].

Over time, buckminsterfullerene (C60) has received significant attention due to its remarkable photochemical and photophysical characteristics. The ability to produce singlet oxygen is an important feature of fullerenes. The use of fullerenes in photodynamic therapy (PDT) can therefore be utilized to treat cancer and kill microorganisms [62]. With hydrophilic groups, fullerene surfaces are easily functionalized, therefore increasing their solubility in solvents like water [67]. The fabrication of molecular entities with functional groups that are covalently linked to the surface of the fullerene cage is made possible by the cage-like structure of fullerenes. This property makes them reliable for drug delivery in targeted areas. In materials science, CBN is now placed at the forefront due to the discovery of graphene, whose thickness is equal to one atom layer of carbon. High intrinsic carrier mobility, optical transmittance, Young's modulus, and thermal conductance are a few of the notable properties of graphene [62].

There are numerous ways to synthesize carbon nanomaterials, which can be classified as top-down and bottom-up processes [68].

3.3.4 NANOCOMPOSITE

A nanocomposite is a unique one-dimensional nanoparticle made up of two or more constituents, leading to the formation of new components. The components are combined such that the resultant nanoparticle exhibits averaged qualities of the two components. The two materials are combined because the combined materials display superior catalytic, thermal stability, and adsorption qualities than the individual components [69, 70]. The two phases, continuous (matrix) and discontinuous (reinforcing material), make up the composite material. Nanocomposites are classified into two types. Functional nanocomposites are classified according to their optical, magnetic, and electrical characteristics, whereas structural nanocomposites are classified according to their mechanical characteristics [49].

3.4 NANOTECHNOLOGICAL APPROACHES FOR IMPROVING SOIL HEALTH

There are different applications of nanotechnology in soil health improvement, as depicted in Figure 3.2, and discussed further in the upcoming sections. Moreover, Table 3.1 also briefly discusses the effect of different types of nanoparticles on soil, along with their performance.

3.4.1 CARBON-BASED NANOPARTICLES

Carbon nanoparticles are commonly employed in tissue engineering, targeted drug delivery, and other commercial applications such as quantum computers, conducting materials, and optical devices [71]. By controlling the release of nutrients and enriching the soil sustainably, carbon-based nanofertilizers in agriculture are more beneficial for improving soil fertility. The applications of carbon-based nanomaterials in agriculture have expanded due to their eco-friendliness and biocompatibility, and also due to their easy surface functionalization. The application of carbon nanomaterials as micronutrients, nanoherbicides, nanopesticides, and composite biofertilizers has been improved due to their specific thermal, mechanical, chemical, and high stability [72, 73].

Fertilizers are required for the production of crops, with nitrogen being the most limiting nutrient. Nitrogen is provided to the plants in the form of urea. However, due to volatilization, about 50% of nitrogen is lost [74]. The low solubility of phosphorus is another major problem in agriculture [75]. Soils have high quantities of phosphates in organic and inorganic forms. However, between 88 and 99% of inorganic phosphorus is bonded to calcium ions and hence inaccessible to plants. For good crop development, over 30% of the agricultural land requires fertilizers to increase the availability of phosphorus [76]. Due to immobilization, phosphate fertilizers create complexes that are insoluble in the soil, therefore, reducing the availability [77]. In a recent study, carbon nanoparticles were able to raise the soil's nitrogen and phosphorus levels, which aided the growth of *Vigna radiata* [78]. *T. aestivum* and *N. tabaccum* subjected to varied proportions of CBNs exhibited similar gains in growth rate, which was depicted by increased nitrogen and phosphorus levels, much higher than the growth attained by using fertilizers [79, 80].

Carbon-based nanoparticles are excellent nanocarriers due to their uniform dispersion, uniform molecular structure, low toxicity, and superior biocompatibility [81]. Carbon-based nanoparticles, like graphene oxide, are excellent carriers of elements when they need to be transported in trace quantities. The presence of oxygen-containing functional groups on the graphene oxide nanoparticles allows them to be good adsorbents of trace elements. Moreover, they have two-phase release properties, demonstrating rapid release early on and subsequently slow, continuous release [82]. It has been observed that Zn and Cu elements can be delivered to the soil more effectively by graphene oxide sheets than by zinc or copper salts [83].

Carbon dots are the emerging star of carbon nanomaterials [84]. They are widely used in agriculture due to their amazing features, which include easy fabrication, photoluminescence, biocompatibility, easy and controlled surface modifications, less toxicity, and good water solubility [85]. In a study, CDs applied to the soil improved the capacity of nodules to fix nitrogen and improved nitrogen and water absorption in soybeans under drought stress. Under drought conditions, CDs were able to improve the nitrogenase activity of soybeans by 8.6%. It also increased the concentration of nitrogen by 18.5% and 14.8% in shoots and roots, respectively. To increase the release of nitrogen from the soil, CDs encouraged the production of root exudates, such as fatty acids, organic acids, and polyketides, and also controlled the activity of beneficial microbial communities, such as *Actinobacteria, Proteobacteria, Gemmatimonadetes,* and *Acidobacteria* [86]. CDs are also considered to be good adsorbents of different metal ions [87]. The availability of carboxyl and hydroxyl functional groups on the surface provides CDs the capacity to adsorb several metal ions, including those crucial for plant growth like Fe_{3+}, K_{+}, Zn_{2+}, Ca_{2+}, and Cu_{2+} [88].

3.4.2 ZnO NANOPARTICLES

Zinc oxide nanoparticles are another type of inorganic nanoparticles with a large surface area and nanoparticle size [89]. Because of their remarkable optical, biocompatibility, mechanical, chemical, electrical, and magnetic properties, they are used in different sectors of agriculture, industry, and the environment. They also possess low toxicity [90, 91]. The behavior, course of events, and possible effects of zinc nanoparticles on agricultural environments are governed by these properties. Depending on their properties and how they interact with organic and inorganic components of the soil, zinc nanoparticles will change physically, biologically, and chemically once they enter the soil [90]. Upon proper contact, when the energy required for attraction is more than the energy required for repulsion, and in the presence of other effects like gravity and Brownian motion, zinc nanoparticles aggregate and agglomerate in the soil environment. Surface properties and particle size are also responsible for their aggregation and agglomeration [92–94]. Aggregation of zinc nanoparticles results in the formation of particle flocks that precipitate in the solution due to gravity. Additionally, as the particle size increases, their mobility in the soil decreases [95]. Therefore, the aggregates frequently stay in the top layers of the soil [96]. Hence, once the spaces are substantially filled with these nanoparticles, the characteristics of the soil matrix, like porosity, permeability, bulk density, texture, shear strength, plasticity, and particulate density, may change [97, 98]. Zinc nanoparticles, owing to their large surface area and nanoporous nature, consistently have a major influence on soil's physical characteristics, even when present in small amounts. As a result, zinc nanoparticles may successfully improve soil structure, moisture content, and water retention. Furthermore, the inclusion of the nanoparticles enhances soil porosity and decreases bulk density, resulting in better conditions for root growth [91, 98].

ZnO nanoparticles have the distinctive qualities to elevate the redox potential of soil. The characteristics and quantities of oxidized and reductive chemicals in the soil affect its redox potential. As a result of soil microorganisms using up the oxygen in the soil during floods, a sufficient amount of oxygen does not get absorbed into the soil. Moreover, certain reducing compounds are also produced and released into the soil, thereby decreasing its redox potential [99]. In the absence of kinetic restrictions, zinc nanoparticles dissolve in the aqueous phase of the soil. This can have a direct or indirect impact on the soil as it may alter the redox potential, ionic strength, conductivity, pH, oxygen demand, nutrient composition, and soil enzymes [99–101]. The dissolution of any nanoparticle occurs when an ion from the particle separates and enters the solution from the electrical double layer [102, 103]. A study found that when zinc nanoparticles were applied to soil, the redox potential increased due to nanoparticle dissolution.

3.4.3 POLYMER NANOPARTICLES

Polymer nanoparticles have a size between 1 and 1,000 nm [104]. They are synthesized in nanocapsule or nanosphere forms [105]. The ability to modify the properties of the polymer nanoparticles allows for the possibility of creating distinct nanoparticles that are best suited for different pollutant kinds and soil conditions. Pollutants like polycyclic aromatic hydrocarbons (PAHs) have a significant impact on the productivity of agricultural soils. Roofing tar, crude oil, creosote, coal tar pitch, and coal are the most common sources of PAHs [106]. The sources of PAHs in soil are due to emissions of harmful fumes from vehicles, fossil fuels, and pyrolysis taking place in industries [107]. They are hydrophobic in nature, and they strongly bind to soil particles. Moreover, PAHs are resistant to microbial degradation once they have sorbed on soils [106]. Due to such contaminants, the agricultural soils are highly affected, leaving them unfit for use. Polymer nanoparticles are great adsorbents [108]. The targeted molecules, for example, pollutants, are adsorbed on the exterior surface of the bulk, which is typically in solid form. This solid portion is eventually enclosed within the particle [109]. A study reported the manufacture of amphiphilic polyurethane (APU) nanoparticles which were employed for removing PAH from the contaminated soils. The precursor chains of the nanoparticles are poly(ethylene glycol)-modified urethane acrylate (PMUA) or polyurethane acrylate anionomer (UAA). These chains can be emulsified and crosslinked in the presence of water. Similar to surfactant micelles, APU particles can improve PAH desorption and transport; however, unlike the active surface parts of micelles, the cross-linked precursor chains in APU particles do not get adsorbed or absorbed in the soil, as these chains are not free. These nanoparticles can be constructed in a way such that they can possess hydrophilic groups on their surface that encourage their movement in the soil. Simultaneously, they can be constructed such that they will have inner hydrophobic regions for better affinity for other PAHs, like phenanthrene (PHEN) [106].

3.4.4 CHITOSAN NANOPARTICLES

Chitosan is an environmentally safe, non-toxic, biodegradable, inexpensive, and non-allergenic biopolymer with several uses in the biomedical, culinary, and agricultural industries [110, 111]. It is more effective than bulk equivalents because of its large surface area and nano size of <100 nm [112]. Some of the methods used to fabricate chitosan nanoparticles include emulsion-based evaporation of solvent, ionotropic gelation method, and microemulsion [113]. Chitosan is a well-known antioxidant. It may readily eliminate free radicals by scavenging them and cause metal ion chelation by donating a proton or an unshared pair of electrons [114]. This property of chitosan can be exploited by synthesizing chitosan nanoparticles and using them to mitigate abiotic stresses. One of the chief abiotic variables threatening plant development and production is salinity, which influences molecular, morphological, and physiological properties [115]. Salinity reduces the capacity of plants to draw water and nutrients from the soil. It also promotes the build-up of excess Cl_- and Na_+ ions in the tissues [116]. Such salt stresses leave a large portion of soil unfit for cultivation. Chitosan nanoparticles are a promising candidate treatment that has been utilized to treat salt stress [117]. In a recent study, chitosan nanoparticles were employed to reduce salt stress in Catharanthus roseus. The nanoparticles successfully increased salt stress tolerance by reducing oxidative damage. They activated the antioxidant system in the plant by upregulating genes like ORCA3, GS, and MAPK3, therefore causing better alkaloid accumulation and providing protection from salinity stress [112]. However, in this study, chitosan nanoparticles were used as a foliar spray. From this study, we can hypothesize the use of chitosan nanoparticles in soils affected by salt stress. Further experiments will be required to utilize chitosan nanoparticles for soil health improvement. In another study, chitosan nanoparticles were used for salt stress mitigation in mung beans. Positive results were demonstrated by reduced content of oxidative stress markers like MDA and H_2O_2 reduction [118].

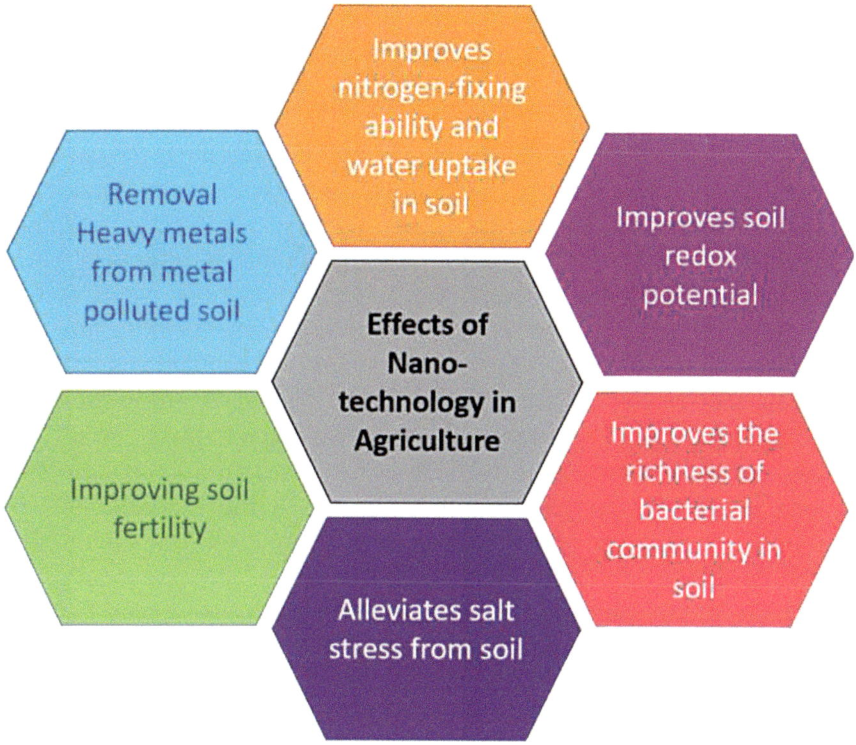

FIGURE 3.2 Nanotechnological approaches for improving soil health.

3.4.5 NANO-BIOCHAR

In recent years, biochar (BC) has gained popularity in the agriculture sector where it is being used to supplement soils. Biochar is produced from biomasses thermochemically, either in the restricted availability of oxygen or no oxygen at all. Biochar has widely been used in different sectors like bioengineering, agriculture, agribusiness, and environmental remediation. Biochar in its bulk form is frequently used in agricultural and environmental applications [119]. Biochar is a thick organic material made by pyrolyzing the biomass of animals or plants [120]. With the evolution of nanotechnology, studies on the sustainable creation of nano-BC for soil applications have been done [119]. The distinction between bulk and nano-BC occurs due to changes in their structure as well as physicochemical properties. The constituent compositions, polar/aromatic nature, pore properties, cationic exchange capacity, pH, specific surface area, and zeta potential of nano-BC can all vary [121]. Nano-BC is predicted to have more desirable features compared to bulk ones due to its small size, differing mobilization, and bioavailability [122].

Heavy metals are regarded as eminent environmental toxins since they are extremely poisonous in nature. Heavy metals enter the environment through volcanic eruptions, the breakdown of metal-containing rocks, and anthropogenic activities, such as mineral extraction, mining, and many more [166]. Over time, heavy metal concentrations in water bodies have steadily increased [123]. Heavy metals in water bodies are difficult to decompose and have strong concealment, leading to long-term toxicity [124]. When such heavy metal-polluted waters enter the soil, they render the area unfit for cultivation [125, 126]. There are several approaches for removing metals from contaminated soils. The majority of treatments, such as flushing and washing of soils, electrokinetics, vapor extraction, solidification, as well as vitrification, are not only costly but upon application do not make the soil suitable for agriculture [127].

These issues, notably in soil, might be addressed by using nano-biochar, which reduces the availability of such contaminants and restores the health of the polluted soils [128, 129]. A study employed nano-biochar in soils that are contaminated with cadmium. The features of nano-biochar that were responsible for lowering cadmium availability in soils were metal ions binding to surface ligands, ionic exchange, and electrostatic interactions. The findings were validated by comparing the concentration of cadmium in the soils of *Brassica chinensis* L. treated with nano-biochar, with untreated soils. Upon the application of nano-biochar, the treated soils displayed a lower concentration of cadmium [130].

In addition to the removal of heavy metals, nano-biochar can also improve the microbial community of the degraded soils. Microorganisms are useful as they increase the availability of soil-borne nutrients [131]. Nano-biochar enhances the soil habitat for microorganisms by increasing water retention, porosity, and tensile strength [132]. Nano-biochar treatment in the soils that are contaminated with cadmium was not only able to minimize the availability of the metal but also enhanced the bacterial population makeup. The addition of nano-biochar increased the diversity of Bacteroidetes and Actinobacteria [130, 133–136, 138].

TABLE 3.1 Effects of Different Types of Nanoparticles

SL. No.	Type of Nanoparticle	Effect of Nanoparticle	Performance	References
1.	Zinc oxide nanoparticle.	Improved soil redox potential.	The soil redox potential was increased by 20~30 mV.	[99]
2.	Graphene oxide nanoparticles.	Removal of As, Cd, Cu, Zn, and Pb	Effectively impacted the availability of metals by immobilizing them.	[139]
3.	Carbon nanoparticle	Improving soil fertility.	The amendment at a concentration of 200 mg/kg$_{-1}$ considerably improved the growth of corn as measured by the increased amount of uptake of nutrients, due to improvement in the availability of P and N in soils.	[140]
4.	Graphene oxide nanoparticle	Improves the availability of zinc and copper better than the respective salts, achieved by controlled release.	A biphasic nutrient release property of the GO-based carrier allowed it to provide micronutrients in both fast-release and slow-release forms.	[141]
5.	Carbon dots	Improves nitrogen-fixing ability, nitrogen availability, and water uptake.	Carbon dots promoted the secretion of fatty acids, organic acids, and polyketides from the roots of the soybean. This enhanced nitrogen release from soil. The nitrogen content was increased by 18.5% in shoots and 14.8% in roots of soybeans.	[86]
6.	Chitosan nanoparticles	Alleviates salt stress	The oxidative stress was reduced by the chitosan nanoparticles, as evidenced by lower levels of H_2O_2 and malondialdehyde. The expression of the genes, GS, MAPK3, and ORCA3 was upregulated by the nanoparticles.	[112]
7.	Iron nanoparticle	Removal of chromium (Cr) and polychlorinated biphenyls (PCB)	The iron nanoparticles efficiently adsorbed Cr and decreased its concentration by 98%. For PCBs, the concentration was reduced by 68%.	[141]
8.	Nano-biochar	Improves the richness of the bacterial community.	Increased the diversity of actinobacteria and bacteroidetes.	[130]
9.	IPMN rice-husk nano-biochar	Arsenic [As (III)] removal	Over 90% of As was eliminated.	[142]
10.	Magnetic biochar derived from wheat straw.	Removal of Hg and antibiotic tetracycline.	Adsorption capacity of 127.4 mg/g for Hg and 268.3 for the antibiotic tetracycline.	[143]

3.5 CHALLENGES

Introducing nanoparticles artificially into the soil could have a considerable impact due to their non-biodegradability and moreover, the tendency to aggregate in soils [167]. Many minuscule soil characteristics appear to be affected by nanoparticles [144]. The maintenance of soil flora is one of the key problems in the subject of soil sustainability [145]. Nanoparticle concentration, soil type, and enzymatic activity of soil all influence the effect of nanoparticles on soil. Josko and associates discovered a detrimental influence on dehydrogenase activity at high concentrations of NPs [146]. The self-cleaning capability of soil and the imbalance of nutrients is the foundation of plant growth. However, the detrimental consequence of NPs affects them highly [147, 148]. Many NPs, particularly those containing zinc, copper, and silver, exhibit antibacterial properties [149]. While such nanoparticles may be useful, their introduction into the environment may endanger valuable microbial ecosystems. The existence of nanoparticles in soil caused significant alterations in carbon fixation, nutrient cycling, and microbially assisted breakdown of various biomolecules, like lipids, and proteins, as well as methane production [168]. Because of their low dissolution, several metal-based nanoparticles are expected to remain in the environment, accumulating in soil and water [150]. Nanoparticles partially dissolve and form aggregates and agglomerates. TiO_2 nanoparticles are very stable after aggregate formation in soil and water, as seen from kinetic studies [151]. Bioaccumulation of such stable nanoparticles may develop because they are easily sorbed. The occurrence of biomagnification also becomes unavoidable within organisms as they interact with different biomolecules, like carbohydrates and lipids, which may then transfer from one organism to the next through different trophic levels [92, 152]. Some metal-based nanoparticles are thermodynamically unstable and dissolve chemically in the absence of kinetic constraints. The dissolution of such nanoparticles produces ionic species that may be hazardous. To further comprehend the possible nanoparticle impacts on organisms over time, the degree to which dissolution will take place and the toxic effects of the same must be evaluated. Metal oxide dissolving behavior in soils may therefore be beneficial in prioritizing nanoparticles for examination and determining which soil variables to take into account that may influence their rate of dissolution [133–136, 138, 153–164].

3.6 CONCLUSION

Soil health is the cornerstone of productive agriculture. Healthy soil is considered to be "living," teaming up with microscopic and larger organisms to perform key functions such as transforming dead matter and minerals into plant nutrients, increasing water retention capacity, improving soil porosity, and enhancing nutrient uptake. These soil properties directly impact crop growth. However, climatic factors such as excess salinity and drought, as well as anthropogenic factors like the irrational use of fertilizers and contamination by heavy metals, can significantly impact soil health, rendering large portions of land unsuitable for cultivation. Heavy metals can also enter the soil through volcanic eruptions and the breakdown of metal-containing rocks. Volatilization or low solubility of essential nutrients like nitrogen and phosphorus can lead to nutrient availability issues, making soils less fertile. Various physical and chemical approaches can help mitigate the effects of climatic and human activities, with the use of fertilizers being a common strategy. The problem of unavailability of nutrients due to volatilization can be solved chemically by using fertilizers. This approach, no doubt, can improve soil health, but only to an extent, after which they become detrimental. Similarly, other techniques, like soil washing and flushing, vitrification, and mulching, are also employed. However, through these techniques, the soil later on becomes unsuitable for agriculture. The nanotechnological approach to improving soil health is a sustainable way of solving the issues discussed above. A variety of nanoparticles exist, such as inorganic, organic, carbon, and nanocomposite-based, that are synthesized by different cost-effective and easy techniques. Nanoparticles hold wide applications in improving soil health by adsorbing harmful heavy metals and polycyclic aromatic hydrocarbons, increasing fertility, mitigating abiotic factors like salt stress, improving the richness of microorganisms in the soil, improving soil redox potential, porosity, water retention capacity, uptake of nutrients, improving the capacity of nodules to fix nitrogen, and improving nitrogenase activity under drought conditions. In addition to such impressive functions of nanoparticles, they also come with some major challenges that need attention in order to make potential use of nanoparticles in improving soil health.

ACKNOWLEDGEMENT

The research was financially supported by the Russian Science Foundation, project no. 21-77-20089

KEYWORDS

- **carbon nanofibers**
- **carbon nanotubes**
- **chemical vapor deposition**
- **nano-biochar**
- **nanoparticles**
- **nanosphere lithography**
- **photodynamic therapy**

REFERENCES

1. Parikh, S. J., & James, B. R. (2012). Soil: The foundation of agriculture. *Nature Education Knowledge, 3*(10), 2.
2. Kumarasinghe, U. (2021). A review of new technologies in soil erosion management. *Journal of Research Technology and Engineering, 2*(1), 120–127.
3. Tilman, D. (1999). Global environmental impacts of agricultural expansion: The need for sustainable and efficient practices. *Proceedings of the National Academy of Sciences, 96*(11), 5995–6000.
4. Pimentel, D., Harvey, C., Resosudarmo, P., Sinclair, K., Kurz, D., McNair, M., & Blair, R. (1995). Environmental and economic costs of soil erosion and conservation benefits. *Science, 267*(5201), 1117–1123.
5. Brodt, S., Six, J., Feenstra, G., Ingels, C. & Campbell, D. (2011) Sustainable Agriculture. *Nature Education Knowledge 3*(10), 1.
6. Bakshi, M., & Abhilash, P. C. (2020). Nanotechnology for soil remediation: Revitalizing the tarnished resource. In *Nano-Materials as Photocatalysts for Degradation of Environmental Pollutants* (pp. 345–370). Elsevier.
7. Corsi, I., Winther-Nielsen, M., Sethi, R., Punta, C., Della Torre, C., Libralato, G., & Buttino, I. (2018). Ecofriendly nanotechnologies and nanomaterials for environmental applications: Key issue and consensus recommendations for sustainable and eco-safe nanoremediation. *Ecotoxicology and Environmental Safety, 154*, 237–244.

8. Guerra, F. D., Attia, M. F., Whitehead, D. C., & Alexis, F. (2018). Nanotechnology for environmental remediation: Materials and applications. *Molecules, 23*(7), 1760.
9. Alazaiza, M. Y., Albahnasawi, A., Ali, G. A., Bashir, M. J., Copty, N. K., Amr, S. S. A., & Al Maskari, T. (2021). Recent advances of nanoremediation technologies for soil and groundwater remediation: A review. *Water, 13*(16), 2186.
10. Meena, M., Zehra, A., Swapnil, P., Harish, Marwal, A., Yadav, G., & Sonigra, P. (2021). Endophytic nanotechnology: An approach to study scope and potential applications. *Frontiers in Chemistry, 9*, 613343.
11. Ramanathan, S., Gopinath, S. C., Arshad, M. M., Poopalan, P., & Perumal, V. (2021). Nanoparticle synthetic methods: Strength and limitations. In *Nanoparticles in Analytical and Medical Devices* (pp. 31–43). Elsevier.
12. Khan, I., Saeed, K., & Khan, I. (2019). Nanoparticles: Properties, applications and toxicities. *Arabian Journal of Chemistry, 12*(7), 908–931.
13. Iravani, S. (2011). Green synthesis of metal nanoparticles using plants. *Green Chemistry, 13*(10), 2638–2650.
14. Ijaz, I., Gilani, E., Nazir, A., & Bukhari, A. (2020). Detail review on chemical, physical, and green synthesis, classification, characterizations, and applications of nanoparticles. *Green Chemistry Letters and Reviews, 13*(3), 223–245.
15. Salavati-Niasari, M., Davar, F., & Mir, N. (2008). Synthesis and characterization of metallic copper nanoparticles via thermal decomposition. *Polyhedron, 27*(17), 3514–3518.
16. Kelly, C. H., & Lein, M. (2016). Choosing the right precursor for thermal decomposition solution-phase synthesis of iron nanoparticles: Tunable dissociation energies of ferrocene derivatives. *Physical Chemistry Chemical Physics, 18*(47), 32448–32457.
17. Ahab, A., Rohman, F., Iskandar, F., Haryanto, F., & Arif, I. (2016). A simple straightforward thermal decomposition synthesis of PEG-covered Gd_2O_3 (Gd_2O_3@PEG) nanoparticles. *Advanced Powder Technology, 27*(4), 1800–1805.
18. Janot, R., & Guérard, D. (2005). Ball-milling in liquid media: Applications to the preparation of anodic materials for lithium-ion batteries. *Progress in Materials Science, 50*(1), 1–92.
19. Koch, C. C., & Whittenberger, J. D. (1996). Mechanical milling/alloying of intermetallics. *Intermetallics, 4*(5), 339–355.
20. Tjong, S. C., & Chen, H. (2004). Nanocrystalline materials and coatings. *Materials Science and Engineering: R: Reports, 45*(1–2), 1–88.
21. Benjamin, J. S. (1990). Mechanical alloying—A perspective. *Metal Powder Report, 45*(2), 122–127.
22. Yadav, T. P., Yadav, R. M., & Singh, D. P. (2012). Mechanical milling: A top-down approach for the synthesis of nanomaterials and nanocomposites. *Nanoscience and Nanotechnology, 2*(3), 22–48.
23. Nandi, P., Chattopadhyay, P. P., Pabi, S. K., & Manna, I. (2003). Solid state synthesis of Al-based amorphous and nanocrystalline Al–Cu–Nb alloys. *Materials Science and Engineering: A, 359*(1–2), 11–17.
24. Khakbiz, M., & Akhlaghi, F. (2009). Synthesis and structural characterization of Al–B_4C nano-composite powders by mechanical alloying. *Journal of Alloys and Compounds, 479*(1–2), 334–341.

25. Colson, P., Henrist, C., & Cloots, R. (2013). Nanosphere lithography: A powerful method for the controlled manufacturing of nanomaterials. *Journal of Nanomaterials, 2013*, 21–21.
26. Pimpin, A., & Srituravanich, W. (2012). Review on micro-and nanolithography techniques and their applications. *Engineering Journal, 16*(1), 37–56.
27. Kim, M., Osone, S., Kim, T., Higashi, H., & Seto, T. (2017). Synthesis of nanoparticles by laser ablation: A review. *KONA Powder and Particle Journal, 34*, 80–90.
28. Yang, G. W. (2007). Laser ablation in liquids: Applications in the synthesis of nanocrystals. *Progress in Materials Science, 52*(4), 648–698.
29. Hahn, A., Barcikowski, S., & Chichkov, B. N. (2008). Influences on nanoparticle production during pulsed laser ablation. *Pulse, 40*(45), 50.
30. Ayyub, P., Chandra, R., Taneja, P., Sharma, A. K., & Pinto, R. (2001). Synthesis of nanocrystalline material by sputtering and laser ablation at low temperatures. *Applied Physics A, 73*, 67–73.
31. Wender, H., Migowski, P., Feil, A. F., Teixeira, S. R., & Dupont, J. (2013). Sputtering deposition of nanoparticles onto liquid substrates: Recent advances and future trends. *Coordination Chemistry Reviews, 257*(17–18), 2468–2483.
32. Muñoz-García, L., Vázquez, R., Cuerno, J., Sánchez-García, M., Castro, M., & Gago, R. (2009). Toward functional nanomaterials. In *Springer US* (pp. 323–398). New York, NY.
33. Lugscheider, E., Bärwulf, S., Barimani, C., Riester, M., & Hilgers, H. (1998). Magnetron-sputtered hard material coatings on thermoplastic polymers for clean room applications. *Surface and Coatings Technology, 108*, 398–402.
34. Nie, M., Sun, K., & Meng, D. D. (2009). Formation of metal nanoparticles by short-distance sputter deposition in a reactive ion etching chamber. *Journal of Applied Physics, 106*(5), 054314.
35. Kumar, S., Bhushan, P., & Bhattacharya, S. (2018). Fabrication of nanostructures with a bottom-up approach and their utility in diagnostics, therapeutics, and others. In *Environmental, Chemical and Medical Sensors* (pp. 167–198).
36. Jones, A. C., & Hitchman, M. L. (2009). Overview of chemical vapor deposition. In *Chemical Vapor Deposition: Precursors, Processes, and Applications* (pp. 1–36).
37. Shah, K. A., & Tali, B. A. (2016). Synthesis of carbon nanotubes by catalytic chemical vapor deposition: A review on carbon sources, catalysts, and substrates. *Materials Science in Semiconductor Processing, 41*, 67–82.
38. Ago, H. (2015). CVD growth of high-quality single-layer graphene. In *Frontiers of Graphene and Carbon Nanotubes: Devices and Applications* (pp. 3–20).
39. Danks, A. E., Hall, S. R., & Schnepp, Z. (2016). The evolution of 'sol-gel' chemistry as a technique for materials synthesis. *Materials Horizons, 3*(2), 91–112.
40. Kumar, D. S., Kumar, B. J., & Mahesh, H. M. (2018). Quantum nanostructures (QDs): An overview. In *Synthesis of Inorganic Nanomaterials* (pp. 59–88).
41. de Coelho Escobar, C., & dos Santos, J. H. Z. (2014). Effect of the sol-gel route on the textural characteristics of silica imprinted with Rhodamine B. *Journal of Separation Science, 37*(7), 868–875.
42. Parashar, M., Shukla, V. K., & Singh, R. (2020). Metal oxide nanoparticles via sol-gel method: A review on synthesis, characterization, and applications. *Journal of Materials Science: Materials in Electronics, 31*, 3729–3749.

43. Zahid, M. U., Pervaiz, E., Hussain, A., Shahzad, M. I., & Niazi, M. B. K. (2018). Synthesis of carbon nanomaterials from different pyrolysis techniques: A review. *Materials Research Express, 5*(5), 052002.
44. Singh, P., Kim, Y. J., Zhang, D., & Yang, D. C. (2016). Biological synthesis of nanoparticles from plants and microorganisms. *Trends in Biotechnology, 34*(7), 588–599.
45. Duan, H., Wang, D., & Li, Y. (2015). Green chemistry for nanoparticle synthesis. *Chemical Society Reviews, 44*(16), 5778–5792.
46. Kathiresan, K., Manivannan, S., Nabeel, M. A., & Dhivya, B. (2009). Studies on silver nanoparticles synthesized by a marine fungus, *Penicillium fellutanum* isolated from coastal mangrove sediment. *Colloids and Surfaces B: Biointerfaces, 71*(1), 133–137.
47. Makarov, V. V., Love, A. J., Sinitsyna, O. V., Makarova, S. S., Yaminsky, I. V., Taliansky, M. E., & Kalinina, N. O. (2014). "Green" nanotechnologies: Synthesis of metal nanoparticles using plants. *Acta Naturae, 6*(1), 35–44.
48. Toshima, N., & Yonezawa, T. (1998). Bimetallic nanoparticles—Novel materials for chemical and physical applications. *New Journal of Chemistry, 22*(11), 1179–1201.
49. Sharma, G., Kumar, A., Sharma, S., Naushad, M., Dwivedi, R. P., ALOthman, Z. A., & Mola, G. T. (2019). Novel development of nanoparticles to bimetallic nanoparticles and their composites: A review. *Journal of King Saud University-Science, 31*(2), 257–269.
50. Auffan, M., Rose, J., Bottero, J. Y., Lowry, G. V., Jolivet, J. P., & Wiesner, M. R. (2009). Towards a definition of inorganic nanoparticles from an environmental, health, and safety perspective. *Nature Nanotechnology, 4*(10), 634–641.
51. Ealia, S. A. M., & Saravanakumar, M. P. (2017). A review of the classification, characterization, and synthesis of nanoparticles and their application. In *IOP Conference Series: Materials Science and Engineering* (Vol. 263, No. 3, p. 032019). IOP Publishing.
52. Mody, V. V., Siwale, R., Singh, A., & Mody, H. R. (2010). Introduction to metallic nanoparticles. *Journal of Pharmacy and Bio Allied Sciences, 2*(4), 282.
53. Fedlheim, D. L., & Foss, C. A. (2001). *Metal Nanoparticles: Synthesis, Characterization, and Applications* (1st ed.). CRC Press. https://doi.org/10.1201/9780367800475.
54. Tai, C. Y., Tai, C. T., Chang, M. H., & Liu, H. S. (2007). Synthesis of magnesium hydroxide and oxide nanoparticles using a spinning disk reactor. *Industrial & Engineering Chemistry Research, 46*(17), 5536–5541.
55. Wijesinghe, W. P. S. L., Mantilaka, M. M. M. G. P. G., Ruparathna, K. A. A., Rajapakshe, R. B. S. D., Sameera, S. A. L., & Thilakarathna, M. G. G. S. N. (2020). Filler matrix interfaces of inorganic/biopolymer composites and their applications. *Interfaces in Particle and Fiber Reinforced Composites*, 95–112. https://doi.org/10.1016/B978-0-08-102665-6.00004-2.
56. Kumar, R., & Lal, S. (2014). Synthesis of organic nanoparticles and their applications in drug delivery and food nanotechnology: A review. *Journal of Nanomaterials & Molecular Nanotechnology, 3*(4), 1–11.
57. Pan, K., & Zhong, Q. (2016). Organic nanoparticles in foods: Fabrication, characterization, and utilization. *Annual Review of Food Science and Technology, 7*, 245–266.

58. Asahi, T., Sugiyama, T., & Masuhara, H. (2008). Laser fabrication and spectroscopy of organic nanoparticles. *Accounts of Chemical Research, 41*(12), 1790–1798.
59. Romero, G., & Moya, S. E. (2012). Synthesis of organic nanoparticles. In *Frontiers of Nanoscience* (Vol. 4, pp. 115–141). Elsevier.
60. Joudeh, N., & Linke, D. (2022). Nanoparticle classification, physicochemical properties, characterization, and applications: A comprehensive review for biologists. *Journal of Nanobiotechnology, 20*(1), 262.
61. Konios, D., Stylianakis, M. M., Stratakis, E., & Kymakis, E. (2014). Dispersion behavior of graphene oxide and reduced graphene oxide. *Journal of Colloid and Interface Science, 430*, 108–112.
62. Patel, K. D., Singh, R. K., & Kim, H. W. (2019). Carbon-based nanomaterials as an emerging platform for theranostics. *Materials Horizons, 6*(3), 434–469.
63. Ruoff, R. S., & Lorents, D. C. (1995). Mechanical and thermal properties of carbon nanotubes. *Carbon, 33*(7), 925–930.
64. Collins, P. G., Bradley, K., Ishigami, M., & Zettl, D. A. (2000). Extreme oxygen sensitivity of electronic properties of carbon nanotubes. *Science, 287*(5459), 1801–1804.
65. Yang, S. Y., Lin, W. N., Huang, Y. L., Tien, H. W., Wang, J. Y., Ma, C. C. M., Wang, Y. S. (2011). Synergetic effects of graphene platelets and carbon nanotubes on the mechanical and thermal properties of epoxy composites. *Carbon, 49*(3), 793–803.
66. Hamada, N., Sawada, S. I., & Oshiyama, A. (1992). New one-dimensional conductors: Graphitic microtubules. *Physical Review Letters, 68*(10), 1579.
67. Herreros-López, A., Carini, M., Da Ros, T., Carofiglio, T., Marega, C., La Parola, V., Prato, M. (2017). Nanocrystalline cellulose-fullerene: Novel conjugates. *Carbohydrate Polymers, 164*, 92–101.
68. Crevillen, A. G., Escarpa, A., & García, C. D. (2018). Carbon-based Nanomaterials in Analytical Chemistry, C. D. Garcia, A. G. Crevillén, & A. Escarpa, (ed.). The Royal Society of Chemistry, 2018, ch. 1, pp. 1-36. https://doi.org/10.1039/9781788012751-00001.
69. Sharma, G., ALOthman, Z. A., Kumar, A., Sharma, S., Ponnusamy, S. K., & Naushad, M. (2017). Fabrication and characterization of a nanocomposite hydrogel for combined photocatalytic degradation of a mixture of malachite green and fast green dye. *Nanotechnology for Environmental Engineering, 2*, 1–7.
70. Pathania, D., Sharma, G., & Thakur, R. (2015). Pectin@ zirconium (IV) silicophosphate nanocomposite ion exchanger: Photocatalysis, heavy metal separation and antibacterial activity. *Chemical Engineering Journal, 267*, 235–244.
71. Husen, A., & Siddiqi, K. S. (2014). Carbon and fullerene nanomaterials in plant system. *Journal of Nanobiotechnology, 12*(1), 1–10.
72. Duhan, J. S., Kumar, R., Kumar, N., Kaur, P., Nehra, K., & Duhan, S. (2017). Nanotechnology: The new perspective in precision agriculture. *Biotechnology Reports, 15*, 11–23.
73. Aacharya, R., & Chhipa, H. (2020). Nanocarbon fertilizers: Implications of carbon nanomaterials in sustainable agriculture production. In *Carbon Nanomaterials for Agri-Food and Environmental Applications* (pp. 297–321). Elsevier.
74. Giroto, A. S., do Valle, S. F., Guimarães, G. G., Wuyts, N., Ohrem, B., Jablonowski, N. D., & Mattoso, L. H. C. (2022). Zinc loading in urea-formaldehyde nanocomposites

increases nitrogen and zinc micronutrient fertilization efficiencies in the poor sand substrate. *Science of the Total Environment, 841*, 156688.
75. Scholz, R. W., Roy, A. H., Hellums, D. T., Roy, A. H., Hellums, D. T., Scholz, R. W., ... & Scholz, R. W. (2014). Sustainable phosphorus management: a transdisciplinary challenge. In *Sustainable phosphorus management: A global transdisciplinary roadmap* (pp. 1–128). Dordrecht: Springer Netherlands. https://doi.org/10.1007/978-94-007-7250-2.
76. Vance, C. P., Uhde-Stone, C., & Allan, D. L. (2003). Phosphorus acquisition and use: Critical adaptations by plants for securing a nonrenewable resource. *New Phytologist, 157*(3), 423–447.
77. Vassilev, N., & Vassileva, M. (2003). Biotechnological solubilization of rock phosphate on media containing agro-industrial wastes. *Applied Microbiology and Biotechnology, 61*, 435–440.
78. Shekhawat, G. S., Mahawar, L., Rajput, P., Rajput, V. D., Minkina, T., & Singh, R. K. (2021). Role of engineered carbon nanoparticles (CNPs) in promoting growth and metabolism of *Vigna radiata* (L.) Wilczek: Insights into the biochemical and physiological responses. *Plants, 10*(7), 1317.
79. Saxena, M., Maity, S., & Sarkar, S. (2014). Carbon nanoparticles in 'biochar' boost wheat (*Triticum aestivum*) plant growth. *RSC Advances, 4*(75), 39948–39954.
80. Liang, T., Yin, Q., Zhang, Y., Wang, B., Guo, W., Wang, J., & Xie, J. (2013). Effects of carbon nanoparticles application on the growth, physiological characteristics and nutrient accumulation in tobacco plants. *Journal of Food, Agriculture & Environment, 11*(3/4), 954–958.
81. Zhu, L., Chen, L., Gu, J., Ma, H., & Wu, H. (2022). Carbon-based nanomaterials for sustainable agriculture: Their application as light converters, nanosensors, and delivery tools. *Plants, 11*(4), 511.
82. Zhou, D. D., Jiang, X. H., Lu, Y., Fan, W., Huo, M. X., & Crittenden, J. C. (2016). Cotransport of graphene oxide and Cu (II) through saturated porous media. *Science of the Total Environment, 550*, 717–726.
83. Kabiri, S., Degryse, F., Tran, D. N., da Silva, R. C., McLaughlin, M. J., & Losic, D. (2017). Graphene oxide: A new carrier for slow release of plant micronutrients. *ACS Applied Materials & Interfaces, 9*(49), 43325–43335.
84. Cui, L., Ren, X., Sun, M., Liu, H., & Xia, L. (2021). Carbon dots: Synthesis, properties and applications. *Nanomaterials, 11*(12), 3419.
85. Li, Y., Xu, X., Wu, Y., Zhuang, J., Zhang, X., Zhang, H., Liu, Y. (2020). A review of the effects of carbon dots in plant systems. *Materials Chemistry Frontiers, 4*(2), 437–448.
86. Wang, C., Ji, Y., Cao, X., Yue, L., Chen, F., Li, J., Xing, B. (2022). Carbon dots improve nitrogen bioavailability to promote the growth and nutritional quality of Soybeans under drought stress. *ACS Nano, 16*(8), 12415–12424.
87. Yahaya Pudza, M., Zainal Abidin, Z., Abdul Rashid, S., Md Yasin, F., Noor, A. S. M., & Issa, M. A. (2020). Eco-friendly sustainable fluorescent carbon dots for the adsorption of heavy metal ions in an aqueous environment. *Nanomaterials, 10*(2), 315.

88. Li, H., Huang, J., Liu, Y., Lu, F., Zhong, J., Wang, Y., Kang, Z. (2019). Enhanced RuBisCO activity and promoted dicotyledons growth with degradable carbon dots. *Nano Research, 12*, 1585–1593.
89. Chaudhuri, S. K., & Malodia, L. (2017). Biosynthesis of zinc oxide nanoparticles using leaf extract of Calotropis gigantea: Characterization and its evaluation on tree seedling growth in the nursery stage. *Applied Nanoscience, 7*(8), 501–512.
90. Sheteiwy, M. S., Shaghaleh, H., Hamoud, Y. A., Holford, P., Shao, H., Qi, W., Wu, T. (2021). Zinc oxide nanoparticles: Potential effects on soil properties, crop production, food processing, and food quality. *Environmental Science and Pollution Research, 28*(28), 36942–36966.
91. Parihar, V., Raja, M., & Paulose, R. (2018). A brief review of structural, electrical, and electrochemical properties of zinc oxide nanoparticles. *Reviews on Advanced Materials Science, 53*(2), 119–130.
92. Tourinho, P. S., Van Gestel, C. A., Lofts, S., Svendsen, C., Soares, A. M., & Loureiro, S. (2012). Metal-based nanoparticles in soil: Fate, behavior, and effects on soil invertebrates. *Environmental Toxicology and Chemistry, 31*(8), 1679–1692.
93. Sharma, K., Kumar, M., & Bhalla, V. (2015). Aggregates of the pentacene-quinone derivative as reactors for the preparation of Ag@ Cu_2O core-shell NPs: An active photocatalyst for Suzuki and Suzuki type coupling reactions. *Chemical Communications, 51*(63), 12529–12532.
94. Rajput, V. D., Minkina, T. M., Behal, A., Sushkova, S. N., Mandzhieva, S., Singh, R., Movsesyan, H. S. (2018). Effects of zinc-oxide nanoparticles on soil, plants, animals and soil organisms: A review. *Environmental Nanotechnology, Monitoring & Management, 9*, 76–84.
95. Darlington, T. K., Neigh, A. M., Spencer, M. T., Guyen, O. T., & Oldenburg, S. J. (2009). Nanoparticle characteristics affect environmental fate and transport through the soil. *Environmental Toxicology and Chemistry: An International Journal, 28*(6), 1191–1199.
96. Fang, J., Shan, X. Q., Wen, B., Lin, J. M., & Owens, G. (2009). Stability of titania nanoparticles in soil suspensions and transport in saturated homogeneous soil columns. *Environmental Pollution, 157*(4), 1101–1109.
97. Taha, M. R., & Taha, O. M. E. (2012). Influence of nano-material on the expansive and shrinkage soil behavior. *Journal of Nanoparticle Research, 14*, 1–13.
98. Bayat, H., Kolahchi, Z., Valaey, S., Rastgou, M., & Mahdavi, S. (2018). Novel impacts of nanoparticles on soil properties: Tensile strength of aggregates and compression characteristics of soil. *Archives of Agronomy and Soil Science, 64*(6), 776–789.
99. Zhang, W., Long, J., Li, J., Zhang, M., Ye, X., Chang, W., & Zeng, H. (2020). Effect of metal oxide nanoparticles on the chemical speciation of heavy metals and micronutrient bioavailability in paddy soil. *International Journal of Environmental Research and Public Health, 17*(7), 2482.
100. Du, W., Sun, Y., Ji, R., Zhu, J., Wu, J., & Guo, H. (2011). TiO_2 and ZnO nanoparticles negatively affect wheat growth and soil enzyme activities in agricultural soil. *Journal of Environmental Monitoring, 13*(4), 822–828.
101. Ge, Y., Priester, J. H., Van De Werfhorst, L. C., Schimel, J. P., & Holden, P. A. (2013). Potential mechanisms and environmental controls of TiO_2 nanoparticle effects on soil bacterial communities. *Environmental Science & Technology, 47*(24), 14411–14417.

102. Borm, P., Klaessig, F. C., Landry, T. D., Moudgil, B., Pauluhn, J., Thomas, K., & Wood, S. (2006). Research strategies for safety evaluation of nanomaterials, part V: Role of dissolution in biological fate and effects of nanoscale particles. *Toxicological Sciences, 90*(1), 23–32.
103. Schaumann, G. E., Philippe, A., Bundschuh, M., Metreveli, G., Klitzke, S., Rakcheev, D., & Vogel, H. J. (2015). Understanding the fate and biological effects of Ag- and TiO_2-nanoparticles in the environment: The quest for advanced analytics and interdisciplinary concepts. *Science of the Total Environment, 535*, 3–19.
104. Zielinska, A., Carreiro, F., Oliveira, A. M., Neves, A., Pires, B., Venkatesh, D. N., et al. (2020). Polymeric nanoparticles: Production, characterization, toxicology and ecotoxicology. *Molecules, 25*(16), 3731.
105. Mansha, M., Khan, I., Ullah, N., & Qurashi, A. (2017). Synthesis, characterization, and visible-light-driven photoelectrochemical hydrogen evolution reaction of carbazole-containing conjugated polymers. *International Journal of Hydrogen Energy, 42*(16), 10952–10961.
106. Tungittiplakorn, W., Lion, L. W., Cohen, C., & Kim, J. Y. (2004). Engineered polymeric nanoparticles for soil remediation. *Environmental Science & Technology, 38*(5), 1605–1610.
107. Roy, D., Jung, W., Kim, J., Lee, M., & Park, J. (2022). Polycyclic aromatic hydrocarbons in soil and human health risk levels for various land-use areas in Ulsan, South Korea. *Frontiers in Environmental Science, 9*, 473.
108. Rivas, B. L., Urbano, B. F., & Sánchez, J. (2018). Water-soluble and insoluble polymers, nanoparticles, nanocomposites, and hybrids with the ability to remove hazardous inorganic pollutants in water. *Frontiers in Chemistry, 6*, 320.
109. Rao, J. P., & Geckeler, K. E. (2011). Polymer nanoparticles: Preparation techniques and size-control parameters. *Progress in Polymer Science, 36*(7), 887–913.
110. Asgari-Targhi, G., Iranbakhsh, A., & Ardebili, Z. O. (2018). Potential benefits and phytotoxicity of bulk and nano-chitosan on the growth, morphogenesis, physiology, and micropropagation of *Capsicum annuum. Plant Physiology and Biochemistry, 127*, 393–402.
111. Arif, Y., Siddiqui, H., & Hayat, S. (2022). Role of chitosan nanoparticles in the regulation of plant physiology under abiotic stress. In Y. Hayat (Ed.), *Sustainable Agriculture Reviews 53: Nanoparticles: A New Tool to Enhance Stress Tolerance* (pp. 399–413). Cham: Springer International Publishing.
112. Hassan, F. A. S., Ali, E., Gaber, A., Fetouh, M. I., & Mazrou, R. (2021). Chitosan nanoparticles effectively combat salinity stress by enhancing antioxidant activity and alkaloid biosynthesis in *Catharanthus roseus* (L.) G. Don. *Plant Physiology and Biochemistry, 162*, 291–300.
113. Mohammed, M. A., Syeda, J. T., Wasan, K. M., & Wasan, E. K. (2017). An overview of chitosan nanoparticles and its application in non-parenteral drug delivery. *Pharmaceutics, 9*(4), 53.
114. Divya, K., & Jisha, M. S. (2018). Chitosan nanoparticles preparation and applications. *Environmental Chemistry Letters, 16*, 101–112.
115. Hassan, F., & Ali, E. (2014). Effects of salt stress on growth, antioxidant enzyme activity, and some other physiological parameters in jojoba (*Simmondsia chinensis* (Link) Schneider) plant. *Australian Journal of Crop Science, 8*(12), 1615–1624.

116. Carillo, P., Cirillo, C., De Micco, V., Arena, C., De Pascale, S., & Rouphael, Y. (2019). Morpho-anatomical, physiological and biochemical adaptive responses to saline water of *Bougainvillea spectabilis* Willd. trained to different canopy shapes. *Agricultural Water Management, 212*, 12–22.
117. Pliankong, P., Suksa-Ard, P., & Wannakrairoj, S. (2018). Chitosan elicitation for enhancing of vincristine and vinblastine accumulation in cell culture of *Catharanthus roseus* (L.) G. Don. *Journal of Agricultural Science, 10*(12), 287–293.
118. Sen, S. K., Chouhan, D., Das, D., Ghosh, R., & Mandal, P. (2020). Improvisation of salinity stress response in mung bean through solid matrix priming with normal and nano-sized chitosan. *International Journal of Biological Macromolecules, 145*, 108–123.
119. Gao, X., & Wu, H. (2014). Aerodynamic properties of biochar particles: Effect of grinding and implications. *Environmental Science & Technology Letters, 1*(1), 60–64.
120. Tomczyk, A., Sokołowska, Z., & Boguta, P. (2020). Biochar physicochemical properties: Pyrolysis temperature and feedstock kind effects. *Reviews in Environmental Science and Bio/Technology, 19*, 191–215.
121. Oleszczuk, P., Ćwikła-Bundyra, W., Bogusz, A., Skwarek, E., & Ok, Y. S. (2016). Characterization of nanoparticles of biochars from different biomass. *Journal of Analytical and Applied Pyrolysis, 121*, 165–172.
122. Rajput, V. D., Singh, A., Singh, V. K., Minkina, T. M., & Sushkova, S. (2021). Impact of nanoparticles on soil resource. In V. D. Rajput (Ed.), *Nanomaterials for Soil Remediation* (pp. 65–85). Elsevier.
123. Xie, Q., & Ren, B. (2022). Pollution and risk assessment of heavy metals in rivers in the antimony capital of Xikuangshan. *Scientific Reports, 12*(1), 14393.
124. Yu, Y., Lü, Y. N., Wang, W. J., Qu, X. D., Liu, J. T., & Wen, C. Y. (2020). Spatio-temporal distribution and risk assessment of heavy metals in middle and lower reaches of Le'an River. *Huan Jing ke Xue = Huanjing Kexue, 41*(2), 691–701.
125. Reddy, V. R., & Behera, B. (2006). Impact of water pollution on rural communities: An economic analysis. *Ecological Economics, 58*(3), 520–537.
126. Srivastava, V., Sarkar, A., Singh, S., Singh, P., De Araujo, A. S., & Singh, R. P. (2017). Agroecological responses of heavy metal pollution with special emphasis on soil health and plant performances. *Frontiers in Environmental Science, 5*, 64.
127. Marques, A. P., Rangel, A. O., & Castro, P. M. (2009). Remediation of heavy metal contaminated soils: Phytoremediation as a potentially promising clean-up technology. *Critical Reviews in Environmental Science and Technology, 39*(8), 622–654.
128. Lu, L., Yu, W., Wang, Y., Zhang, K., Zhu, X., Zhang, Y., & Chen, B. (2020). Application of biochar-based materials in environmental remediation: From multi-level structures to specific devices. *Biochar, 2*, 1–31.
129. Oni, B. A., Oziegbe, O., & Olawole, O. O. (2019). Significance of biochar application to the environment and economy. *Annals of Agricultural Sciences, 64*(2), 222–236.
130. Liu, W., Li, Y., Feng, Y., Qiao, J., Zhao, H., Xie, J., & Liang, S. (2020). The effectiveness of nano biochar for reducing phytotoxicity and improving soil remediation in cadmium-contaminated soil. *Scientific Reports, 10*(1), 858.
131. Jacoby, R., Peukert, M., Succurro, A., Koprivova, A., & Kopriva, S. (2017). The role of soil microorganisms in plant mineral nutrition—Current knowledge and future directions. *Frontiers in Plant Science, 8*, 1617.

132. Ameloot, N., Graber, E. R., Verheijen, F. G., & De Neve, S. (2013). Interactions between biochar stability and soil organisms: Review and research needs. *European Journal of Soil Science, 64*(4), 379–390.
133. Singh, A., Rajput, V. D., Rawat, S., Sharma, R., Kumar Singh, A., S. Tomar, R., & Lobzenko, I. P. (2020). Monitoring soil salinity and recent advances in the mechanism of salinity tolerance in plants. *Biogeosystem Technique, 7(2)*, 66–86.
134. Rajput, P., Rajput, V. D., Singh, A., Sharma, R., Rawat, S., S. Tomar, R., Singh, A., & Minkina, T. (2021). Impact of emerging tools for solving 'hidden hunger' in humans. *Emerging Tools for Sustainable Agriculture and Food Security, 1(1)*, 1730.
135. Rajput, V. D., Minkina, T., Kumari, A., Shende, S. S., Ranjan, A., Faizan, M., & Kızılkaya, R. (2022). A review on nanobioremediation approaches for restoration of contaminated soil. *Eurasian Journal of Soil Science, 11*(1), 43–60.
136. Rajput, V. D., Minkina, T., Feizi, M., Kumari, A., Khan, M., Mandzhieva, S., & Choudhary, R. (2021). Effects of silicon and silicon-based nanoparticles on rhizosphere microbiome, plant stress and growth. *Biology, 10*(8), 791.
137. Nascimento, M. A., Cruz, J. C., Rodrigues, G. D., de Oliveira, A. F., & Lopes, R. P. (2018). Synthesis of polymetallic nanoparticles from spent lithium-ion batteries and application in the removal of reactive blue 4 dye. *Journal of Cleaner Production, 202*, 264–272.
138. Singh, A., Sengar, R. S., Shahi, U. P., Rajput, V. D., Minkina, T., & Ghazaryan, K. A. (2023). Prominent effects of zinc oxide nanoparticles on roots of rice (*Oryza sativa* L.) grown under salinity stress. *Stresses, 3*(1), 33–46.
139. Baragaño, D., Forján, R., Welte, L., & Gallego, J. L. R. (2020). Nanoremediation of As and metals polluted soils by means of graphene oxide nanoparticles. *Scientific Reports, 10*(1), 1–10.
140. Zhao, F., Xin, X., Cao, Y., Su, D., Ji, P., Zhu, Z., & He, Z. (2021). Use of carbon nanoparticles to improve soil fertility, crop growth, and nutrient uptake by corn (*Zea mays* L.). *Nanomaterials, 11*(10), 2717.
141. Gil-Díaz, M., Pérez, R. A., Alonso, J., Miguel, E., Diez-Pascual, S., & Lobo, M. C. (2022). Iron nanoparticles to recover a co-contaminated soil with Cr and PCBs. *Scientific Reports, 12*(1), 3541.
142. Nath, B. K., Chaliha, C., & Kalita, E. (2019). Iron oxide permeated mesoporous rice-husk nano biochar (IPMN) mediated removal of dissolved arsenic (As): Chemometric modeling and adsorption dynamics. *Journal of Environmental Management, 246*, 397–409.
143. Li, R., Zhang, Y., Deng, H., Zhang, Z., Wang, J. J., Shaheen, S. M., & Du, J. (2020). Removing tetracycline and Hg (II) with ball-milled magnetic nano biochar and its potential on polluted irrigation water reclamation. *Journal of Hazardous Materials, 384*, 121095.
144. Ben-Moshe, T., Frenk, S., Dror, I., Minz, D., & Berkowitz, B. (2013). Effects of metal oxide nanoparticles on soil properties. *Chemosphere, 90*(2), 640–646.
145. Torsvik, V., & Øvreås, L. (2002). Microbial diversity and function in soil: From genes to ecosystems. *Current Opinion in Microbiology, 5*(3), 240–245.
146. Josko, I., Oleszczuk, P., & Futa, B. (2014). The effect of inorganic nanoparticles (ZnO, Cr_2O_3, CuO, and Ni) and their bulk counterparts on enzyme activities in different soils. *Geoderma, 232*, 528–537.

147. Janvier, C., Villeneuve, F., Alabouvette, C., Edel-Hermann, V., Mateille, T., & Steinberg, C. (2007). Soil health through soil disease suppression: which strategy from descriptors to indicators? *Soil Biology and Biochemistry, 39*(1), 1–23.
148. Suresh, A. K., Pelletier, D. A., & Doktycz, M. J. (2013). Relating nanomaterial properties and microbial toxicity. *Nanoscale, 5*(2), 463–474.
149. Wang, L., Hu, C., & Shao, L. (2017). The antimicrobial activity of nanoparticles: present situation and prospects for the future. *International Journal of Nanomedicine, 12*, 1227–1249.
150. Baun, A., Hartmann, N. B., Grieger, K., & Kusk, K. O. (2008). Ecotoxicity of engineered nanoparticles to aquatic invertebrates: A brief review and recommendations for future toxicity testing. *Ecotoxicology, 17*, 387–395.
151. French, R. A., Jacobson, A. R., Kim, B., Isley, S. L., Penn, R. L., & Baveye, P. C. (2009). Influence of ionic strength, pH, and cation valence on aggregation kinetics of titanium dioxide nanoparticles. *Environmental Science & Technology, 43*(5), 1354–1359.
152. Unrine, J., Bertsch, P., & Hunyadi, S. (2008). Bioavailability, trophic transfer, and toxicity of manufactured metal and metal oxide nanoparticles in terrestrial environments. In S. S. Sadhukhan (Ed.), *Nanoscience and Nanotechnology: Environmental and Health Impacts* (pp. 345–378). Nova Science Publishers.
153. Rajput, V. D., Minkina, T., Kumari, A., Harish, Singh, V. K., Verma, K. K., & Keswani, C. (2021). Coping with the challenges of abiotic stress in plants: New dimensions in the field application of nanoparticles. *Plants, 10*(6), 1221.
154. Rajput, V., Minkina, T., Sushkova, S., Behal, A., Maksimov, A., Blicharska, E., & Barsova, N. (2020). ZnO and CuO nanoparticles: a threat to soil organisms, plants, and human health. *Environmental Geochemistry and Health, 42*, 147–158.
155. Rajput, V. D., Minkina, T., Sushkova, S., Tsitsuashvili, V., Mandzhieva, S., Gorovtsov, A., & Gromakova, N. (2018). Effect of nanoparticles on crops and soil microbial communities. *Journal of Soils and Sediments, 18*, 2179–2187.
156. Rajput, V., Minkina, T., Fedorenko, A., Sushkova, S., Mandzhieva, S., Lysenko, V., & Ghazaryan, K. (2018). Toxicity of copper oxide nanoparticles on spring barley (*Hordeum sativum* distichum). *Science of the Total Environment, 645*, 1103–1113.
157. Rajput, V. D., Minkina, T., Sushkova, S., Barakhov, A., Singh, S., Mandzhieva, S. S., & Garg, M. C. (2022). A practical evaluation of the integrated role of biochar and nanomaterials in soil remediation processes. *Environmental Geochemistry and Health, 45, 9435–9449.*
158. Rajput, V. D., Faizan, M., Upadhyay, S. K., Kumari, A., Ranjan, A., Sushkova, S., & Deryabkina, I. (2022). Influence of Nanoparticles on the Plant Rhizosphere Microbiome. In T. Minkina & S. S. Mandzhieva (Eds.), *The Role of Nanoparticles in Plant Nutrition under Soil Pollution: Nanoscience in Nutrient Use Efficiency* (pp. 83–102). Springer International Publishing.
159. Rajput, V. D., Singh, A., Minkina, T., Rawat, S., Mandzhieva, S., Sushkova, S., & Upadhyay, S. K. (2021). Nano-enabled products: Challenges and opportunities for sustainable agriculture. *Plants, 10*(12), 2727.
160. Singh, A., Singh, A. K., Rawat, S., Pal, N., Rajput, V. D., Minkina, T., & Tripathi, J. N. (2022). Satellite-Based Quantification of Methane Emissions from Wetlands and Rice Paddies Ecosystems in North and Northeast India. *Hydrobiology, 1*(3), 317–330.

161. Singh, A., Sengar, R. S., Rajput, V. D., Minkina, T., & Singh, R. K. (2022). Zinc oxide nanoparticles improve salt tolerance in rice seedlings by improving physiological and biochemical indices. *Agriculture, 12*(7), 1014.
162. Singh, A., Sengar, R. S., Sharma, R., & Rajput, P. (2021). Nano-Priming Technology for Sustainable Agriculture. *Bio Geosystem Technique, 8*(2). pp 1–14, DOI: 10.13187/bgt.2021.2.79
163. Singh, A., Mehrotra, R., Rajput, V. D., Dmitriev, P., Singh, A. K., Kumar, P., & Singh, A. K. (2022). Geoinformatics, artificial intelligence, sensor technology, big data: emerging modern tools for sustainable agriculture. *Sustainable Agriculture Systems and Technologies*, 295–313.
164. Singh, A., Rajput, V. D., Rawat, S., Sharma, R., Singh, A. K., Kumar, P., & Singh, S. (2022). Geoinformatics and nanotechnological approaches for coping up abiotic and biotic stress in crop plants. *Sustainable Agriculture Systems and Technologies*, 337–359.
165. Nascimento, M. A., Cruz, J. C., dos Reis, M. F., de Carvalho Damasceno, O. I., Reis, E. L., Reis, C., & Lopes, R. P. (2018). Synthesis of polymetallic nanoparticles from printed circuit board waste and application in textile dye remediation. *Journal of Environmental Chemical Engineering, 6*(4), 5580–5586.
166. Kapoor, D., & Singh, M. P. (2021). Heavy metal contamination in water and its possible sources. In M. A. Ali (Ed.), *Heavy Metals in the Environment* (pp. 179–189). Elsevier.
167. Rajput, V. D., Minkina, T., Sushkova, S., Tsitsuashvili, V., Mandzhieva, S., Gorovtsov, A., & Gromakova, N. (2018). Effect of nanoparticles on crops and soil microbial communities. *Journal of Soils and Sediments, 18*, 2179–2187.
168. Rajput, V. D., Minkina, T., Ahmed, B., Singh, V. K., Mandzhieva, S., Sushkova, S., & Wang, B. (2022). Nano-biochar: A novel solution for sustainable agriculture and environmental remediation. *Environmental Research, 210*, 112891.

CHAPTER 4

Nano-Biochar: A Review on Its Applications in Agriculture Soil Productivity

ALOK BHARADWAJ, AKANKSHA SHAKYA, and MANSI GUPTA

Department of Biotechnology, GLA University, Mathura, Uttar Pradesh, India

ABSTRACT

In the present scenario, changes in global temperature usually occur due to the release of greenhouse gases and land-use changes. Population explosion and rapid industrialization are mainly responsible for the release of carbon compounds and other gases into the atmosphere. By employing appropriate technologies, there must be a reduction in the emission of these gases. Such carbon solids known as "biochar," have the capability to be used in industries, agriculture, and other sectors.

This 'biochar' when amended with soil improves the physicochemical properties and solute transfer ability, thus increasing the fertility of the soil. However such biochar molecules, when applied on a nano-scale, have shown conflicting results. Nowadays, the term 'Nano-biochar' is receiving much attention due to its improved physicochemical properties. Recent research is ongoing to find novel technologies for constructing this 'Nano-biochar.' This 'Nano-biochar' is a nano-sized biochar material with better physicochemical properties. It exhibits several characteristics like better plant growth and soil properties, along with minimal chances of plant disease, wastewater treatment, and bioremediation of pesticides and other

Nano-Biochar and Nanoparticles for Soil Health Improvement.
Vishnu D. Rajput and Tatiana M. Minkina (Eds.)

pollutants. Due to its lower cost, better sustainability, and environmentally friendly approach, it has become an alternative to conventional procedures. Additionally, it also minimizes climate change through its carbon sequestration function. Therefore, compared to biochar, nano-biochar shows better potential to absorb pollutants, nutrients, and contaminants. In this review, we discuss various applications of nano-biochar in agriculture and soil management.

4.1 INTRODUCTION

Greenhouse gas emissions are undeniably a major contributor to climate change, and the exact role of each cause is still up for debate. A rise in human activity over recent decades has resulted in the creation and release of these gases [1–5]. One technique for absorbing and segregating carbon [6–10] is to capture greenhouse gases at industrial facilities' output and deposit them in safe storage tanks. The global carbon cycle and ecology are also negatively impacted by CO_2 emissions from the soil [11–13]. The greatest carbon store is soil organic carbon (SOC), which is double the quantity of carbon in the atmosphere [14–17].

Inappropriate cultivation and plowing typically result in a large loss of organic carbon in agricultural soils [18–22]. Changes in plowing methods, sufficient fallow, the introduction of plants with deep roots, and the sequestration of carbon all contribute to a higher organic carbon content in the soil. Carbon sequestration, meanwhile, is carried out in numerous methods [23–28]. The benefits and drawbacks of a few of these strategies are outlined in Table 4.1. According to the literature [29–31], the most suitable approach is the one where the resulting carbon has long-lasting effects and is very stable against chemical and biological agents. Increases in growth rate and yield can be attributed to biochar's use because of its ability to improve the soil's physical and chemical qualities, resulting in a decrease in soil organic matter and a corresponding increase in soil carbon dioxide (Table 4.1) [32–36].

Biochar is a carbonized solid produced by the controlled thermal decomposition of biomass feedstock, such as agricultural waste and other lignocellulosic materials. This process takes place in an oxygen-poor (gasification) or oxygen-deficient environment (pyrolysis). Several thermochemical techniques can turn solid wastes and animal litter into

TABLE 4.1 Various Modes of Carbon Sequestration

Sl. No.	Carbon Sequestration Procedure	Characteristics
1.	Absorption and deposition by plants.	Quite a normal process but carbon accumulates in the atmosphere through plant decomposition.
2.	The plantation is devoid of plowing and crop rotation.	It's a time-consuming process.
3.	Plants with deep roots	It's a normal process but has short-term impacts.
4.	Biochar applications	It's a constant procedure having long-lasting impacts and requires more research.
5.	Geologic sequestration	It's quite expensive process and very complicated technology.

biochar. Scientists are increasingly interested in biochar technology due to its potential to reduce greenhouse gas emissions, cleanse the environment of harmful pollutants, and boost crop yields, soil quality, and agricultural productivity [37–41]. Additionally, since biochar can be generated at temperatures as low as 700°C and requires very little energy, it is not only highly economical and environmentally beneficial but also effective in reusing waste materials. At the same time, feedstock for biochar, such as solid waste and agricultural biomass, is widely available [42–47]. However, nano biochar has evolved as a means of overcoming biochar's limitations by enhancing its physical, chemical, and structural qualities. Its nanoscale size and abundance of functional groups with enhanced mechanical and thermal stability set it apart and provide it an edge, as recently mentioned by Noreen and Abd-Elsalam [48]. With a greater adsorption capacity, micro biochar may generate more reactive organic species (ROS) than macro biochar due to its more negative zeta potential, smaller hydrodynamic radius, and higher number of functional groups containing oxygen and carbon defects [49].

Due to its physical and chemical qualities and porous structure, nano biochar is now being considered as a soil amendment, enzyme support, and excellent adsorbent. The nano biochar particles, with their larger specific surface area, micro-porosity, and surface hydrophobicity, show increased absorption ability for a variety of contaminants, such as herbicides, heavy metals, PCBs, and PAHs, and they might be used to counteract soil acidity because of their higher alkalinity and pH [50, 51]. Since then, nanobiochar has been incorporated into nanocomposites for application in purifying water, generating electricity, and removing pollutants [52–57]. Its porous structure, the extra number of surface functional groups, and larger surface-to-volume ratio make it an effective immobilization-supporting material for enzymes [58]. Nanocomposites have shown remarkable enhancements in a number of features, including pore characteristics, functional groups, and, consequently, catalytic devastation ability; active sites and separation become easy and fast [53, 59].

4.2 PRODUCTION OF BIOCHAR

Carbon-rich nano biochar is produced by the pyrolysis and carbonization of biomass. The presence or lack of oxygen determines whether thermal breakdown or combustion occurs. When biomass is burned with heat and oxygen, it breaks down into gases and carbon dioxide, as well as ash containing metal oxides, silica, and other volatile inorganic components [60, 61]. Carbon dioxide gas is not created during the thermochemical conversion of biomass if oxygen is excluded from the reaction. Biochar is a solid form of carbon. Syngas, which consists of hydrogen, methane, and other hydrocarbon gases, and liquid bio-oil are two further byproducts of thermal decomposition. Depending on the temperature at which the biomass is heated, thermal breakdown might proceed either quickly or slowly [62, 63]. Thermal decomposition proceeds slowly at rates below 100°C/min, with biochar and syngas emerging as the reaction's final byproducts. Biofuels are often a byproduct of rapid thermal degradation, which occurs at temperatures above 1,000°C/min [2].

4.2.1 STRUCTURAL CHARACTERISTICS OF BIOCHAR

Polycyclic aromatic hydrocarbons (biochar) consist of rings composed of 6 carbon atoms. A strong aromatic structure is present in biochar, making it resistant to deterioration from both biological and chemical processes [64]. Biochar is a mixture of carbon, hydrogen, oxygen, and other elements. Depending on the raw material used to produce the biochar, different minerals including nitrogen, phosphorus, and sulfur can also be present. Many different types of functional groups, including hydroxyl, ketone, ester, aldehyde, amino, nitro, and carboxyl, can be found in this carbon compound. But biochars are rich in humic and fulvic organic acids. In contrast, biochars are loaded with beneficial compounds including humic and fulvic acids. Biochars can interact with both organic and inorganic substances. The vascular system of plants has a special order and is composed of strong cell walls, both of which are inherited by the final biochar during the conversion of plant biomass to nano biochar, which is why maintaining the plant's intrinsic and internal structure (phototaxy) is so crucial. There are three distinct pore sizes in nano biochar:

1. Macrospores, having a diameter greater than 50 nm;
2. Mesoporous, with a diameter between 2 and 50 nm; and
3. Microporous, with a diameter less than 2 nm.

Micrometer porosity is important in biochar because it allows air to enter the material and provides a refuge for microorganisms, allowing them to thrive in the biochar despite any conditions they may encounter. Nanobiochar has a substantially larger specific surface area (around 1,500 m_2/g) than clay particles. The parameters of the thermal decomposition process and the kind of biomass have a direct effect on the properties of the resulting nano biochar, including its elemental composition, porosity percentage, particle or pore size, and the presence of degradable hydrocarbons. Ash content in the raw material has an effect on the carbonization process in nano biochar made from lignin. The created nano biochar exhibits higher qualities and attributes as a result of minimizing this amount of ash. Biochar's porosity is diminished by the incorporation of lipid molecules, humic acid, and folic acid [65]. Some samples may melt and become plastic when heated because their cellular structure is composed of primary biomass that is particularly susceptible to heat. This phenomenon can stop biochar from developing non-micrometer pores. The rate at which the temperature rises also influences the melting of the cell structure. Studies show that Nano biochar with the right structure can be generated from pinewood if the heating rate is 20C/s, but at 500°C/s, the cellular structure of the melt is formed and micrometer-sized pores cannot form. The term "biochar activation" refers to a set of physical and chemical activities that increase the specific surface area of the Nano biochar. Nano biochar can be physically activated by heating it to temperatures between 700 and 1,200°C while exposing it to oxidizing gases including ordinary steam, carbon dioxide, or oxygen. Combining biomass with acids, bases, or salts before carbonization is the foundation of the chemical activation approach [66, 67]. The specific surface area produced by chemically activated Nano biochar is nearly identical to that produced by physical methods, despite the lower carbonization temperatures required for the former [2, 3].

4.3 APPLICATION OF BIOCHAR COMPOSITES IN THE ENVIRONMENT

Many researchers have shown that macro-biochar can be beneficial to the environment. Carbon sequestration, energy production, and phytoremediation (waste management, pollutant removal, wastewater treatment, etc.) are just a few of the environmental uses being investigated with nanobiochar today (Figure 4.1). When compared to regular biochar, nano biochar, and biochar treated with nano minerals showed superior adsorption capabilities for the removal of a wide range of pollutants [68, 69]. In addition, biochar can remedy pollutants that may lead to serious environmental and health problems by adsorbing hazardous substances like pesticides and immobilizing metals. Therefore, nanobiochar has been seen as a potentially useful tool for the bioremediation of several pollutants.

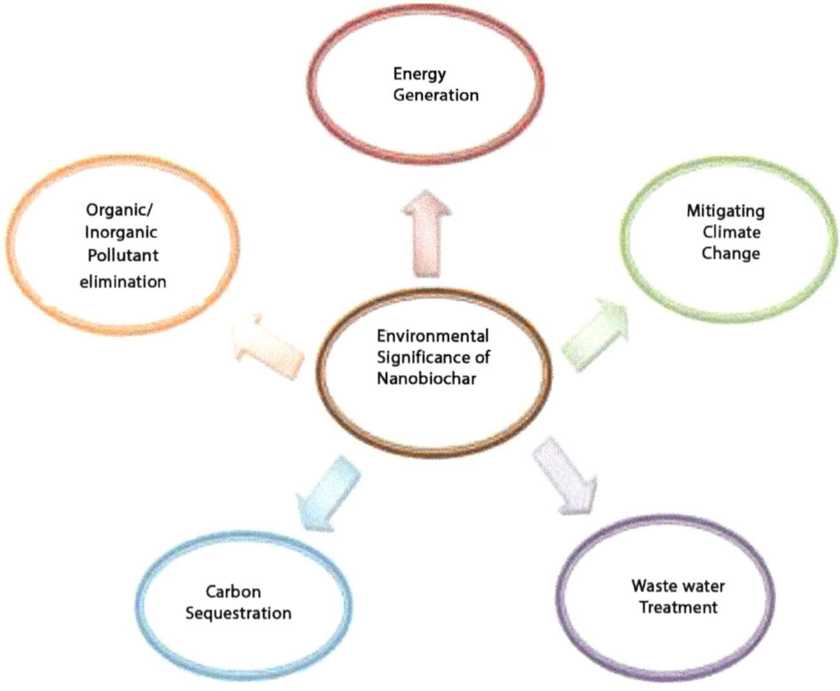

FIGURE 4.1 Environmental significance of nanobiochar composites.

4.3.1 NANO-BIOCHAR AND PHYTOREMEDIATION OF CONTAMINATION

Adsorption is a well-known method for removing contaminants, and biochar has been used for this purpose. The ability to absorb NOM has also been attributed to other nanocarbons such as black, activated, or designed nanomaterial carbon. Researchers have researched biochar's ability to remove contaminants such as heavy metals, pharmaceuticals/drugs, dyes, and more, and several have found nanobiochar to be quite effective in doing so [68, 69]. Mechanical grinding and a top-down technique with dendro biochar from waste material were used to produce nano biochar in the study by Ramanayaka et al. [70], which was then used to remove pollutants such as antibiotics, agrochemicals, and potentially harmful components from aqueous medium.

The results showed that the graphitic nano biochar in dendro biochar was successful in chelating glyphosate (GL), chromium (VI), oxytetracycline (OTC), and cadmium (Cd)(II). The adsorption capacities of ZnO/ZnS modified nano biochar for Cu (II), Pb (II), and Cr (VI) were determined to be 63.29, 27.05, and 15.23 mg/g, respectively [71]. Several further studies have since demonstrated that biochar nanocomposites have a high capacity for removing key heavy metal pollutants such as arsenic, lead, copper, chromium, cadmium, etc., and were also discovered to be effective in eliminating hazardous compounds, colors, medicines, and antibiotics.

4.3.2 NANO-BIOCHAR AND WASTEWATER BIOREMEDIATION

Heavy metals accumulate and become toxic to aquatic life, making heavy metal poisoning a severe environmental problem. Heavy metals cannot be decomposed, in contrast to other organic and inorganic pollutants [59], therefore their removal is an issue. The most widely used heavy metal removal techniques include surface assimilation, ion exchange, chemical precipitation, and others [72], although saturated surface assimilation solutions/emissions have gained popularity because it is economically viable [73]. Adsorbents, such as activated carbon, flat iron oxide, silica gel, zeolite, attapulgite, and carbon-based nanofibers, all have their advantages and disadvantages [59, 74]. Currently, activated charcoal has shown promise as a pollutant adsorption agent, and it may also help reduce water pollution [52]. Nanoparticle-infused biochar has improved qualities and capabilities for a variety of uses, including carbon sequestration and wastewater treatment [59]. The elimination of organic pollutants was greatly aided by the catalytic material-coated charcoal's dual desorption and catalytic degradative actions [53].

The main cause of water contamination was the simultaneous adsorptive and degradative activity of nano biochar and nanocomposites based on biochar on heavy metals such as As, Cr, Pb, Cd, and Cu. According to Li et al. [75], converting pure magnetic biochar (PMBCs) into ball-milled magnetic nano biochar (BMBCs) made from wheat straw significantly improved the removal of tetracycline (TC) and Hg (II) from contaminated irrigation water. Additionally, TC and Hg (II) were 99% absorbed by the action of BMBC within 12 hours. For TC and Hg (II), the removal capacities were 268.3 mg/g and 127.4 mg/g, respectively. Photosynthetic bacteria (PSB) loaded onto Fe_3O_4/biochar nanocomposites were used for wastewater bioremediation in another study by He et al. [76]. This PSB-composite mixture was able to remove 87.5% of NH_{4+}, 82.1% of chemical oxygen demand, and 92.1% of PO_{43} from the wastewater. The PSB/Fe_3O_4/biochar composites were shown to be efficient and applicable for wastewater bioremediation.

4.3.3 NANO-BIOCHAR AND ORGANIC MATTER IN SOILS

As previously established, soil organic carbon (SOC) breaks down over time and releases carbon dioxide gas into the environment. According to the findings, adding Nano biochar to the soil remarkably raises the biological carbon content of the soil [77–79]. As an illustration, when 8% Nano biochar is added to the soil, the amount of biological carbon increases by around 41%, and the Nano biochar remains unchanged after 210 days. According to the findings, Nano biochar is hypothesized to increase soil organic carbon stability and decrease mineralization. The three methods listed below [31–33] can help increase the stability of soil organic carbon:

1. Aggregation-based physical stabilization of organic carbon;
2. Silt or fine-grained soil mixing clay and silt particles with organic carbon; and
3. Formation of biodegradable SOC compounds results in soil biochemical stabilization [78–80].

Among the aforementioned considerations, the incorporation of soil organic carbon into the biochar is most likely to be the cause of the improved SOC stability. The bulk of research shows that adding biochar to soil slows down the decomposition of SOC (Figure 4.1) [1].

4.3.4 SOIL CO_2 EMISSIONS FROM SOIL AND NANO-BIOCHAR

By adding biochar to the soil, carbon dioxide emissions are initially boosted. For instance, adding 8% biochar to the soil increases carbon dioxide emissions for the first 20 days; however, after 120 days, gas emissions are significantly reduced. In other words, increasing soil biochar over time results in a reduction in carbon dioxide emissions. Generally speaking, the physicochemical characteristics of biochar, soil conditions, and soil microbial population all affect how much carbon dioxide is released from the soil [20–22]. The unstable component of biochar, however, may stimulate and promote the growth of microorganisms when added to the soil [81–83]. The soil's ability to hold moisture is improved by the micrometric porosity of biochar, and increased moisture slows the decomposition of SOC. So, by increasing soil moisture, biochar indirectly lowers carbon dioxide emissions (Figure 4.1) [1].

4.3.5 NANO-BIOCHAR AND PHYSICAL PROPERTIES OF SOIL

Tensile strength, fluid mechanics, and gas movement are affected by the physical differences between soil and biochar in the soil. The life of soil creatures may also be impacted by these occurrences. The kind and properties of the biochar being used determine each of these effects. When biochar is added, the soil's tensile strength is reduced when its tensile strength is smaller than the soil's (clay-rich soil). Root growth is significantly impacted by mechanical impedance reduction [23–25]. Despite the fact that this is one of the factors promoting root growth and development in the growth media at the moment, the roots are still more susceptible to soil and the crop can easily produce roots [84–86]. The mobility and displacement of invertebrates in the soil may increase when tensile strength decreases. The movement of invertebrates through the soil and their consumption of materials containing various vitamins and nutrients increase soil fertility. The porosity of the soil, which is raised by biochar by reducing the bulk density of the soil, affects the amount of water maintained in the soil, the model of root elongation, and soil organisms.

4.3.6 NANO-BIOCHAR AND CHEMICAL PROPERTIES OF SOIL

Through processes like aeration, contact with organic matter and soil minerals, and microbial oxidation, the recognized and unidentified organic matter in biochar changes over time. Soil's pH, cation exchange capacity (CEC), electrical conductivity (EC), and EC are only a few of the variables that shift after biochar is introduced. Depending on the starting material, biochar's pH can range anywhere from 4 to 12. Biochar's final pH is affected by the temperature at which it's produced and can change over time [87–90]. Researchers have determined that the oxidation of carbon to carboxylic acid groups is the primary cause of the pH drop. In contrast, it is believed that the dissolution of alkaline substances is responsible for the observed increase in pH [28, 29].

Therefore, the soil's pH may shift as biochar in the soil degrades and reforms over time. Initially having a lower cation exchange capacity than soil organic matter, biochar can have a positive or negative net charge [91–93] (Figure 4.2). The CEC and charge density in the finished product increase with the amount of ash in the raw material used to make the biochar. Of course, the temperature at the time of CEC biochar synthesis also affects how much is produced. The functional groups of organic acids like humic acid are primarily related to the CEC in biochar, and these compounds are destroyed when the formation temperature increases too much. As a result, the biochar's final CEC rate is decreased. The leaching rate of nitrogen fertilizers (ammonium compounds) can be reduced by biochar by roughly 60%, according to a study. Contrarily, this feature results in the adsorption of heavy metals and organic soil pollutants (such as pesticides, etc.), which lowers the soil's contamination level. In addition to increasing its CEC and protecting soil organic carbon, biochar can also absorb soil organic matter.

As a result, the amount of soil organic matter grows over time and the soil organic carbon does not disintegrate [94–96]. Because of its ability to exchange cations with other elements, biochar can effectively reduce the saltiness (EC) and osmotic potential of soil solutions by absorbing substances including sodium, calcium, and magnesium [30–32]. The plant experiences reduced salt stress as a result, and plant output rises. Aluminum and manganese are two elements whose concentration in the soil solution rises in acidic soils [33–35]. In addition to poisoning, the presence of these elements prevents plants from absorbing other nutrients. Biochar can lessen the toxicity of these elements by lowering their concentration. Plants and microorganisms can utilize biochar because certain elements, such as Si, Fe, S, P, K, Mg, and Ca, are released from it and enter the soil solution when it is combined with acidic solutions [2–4].

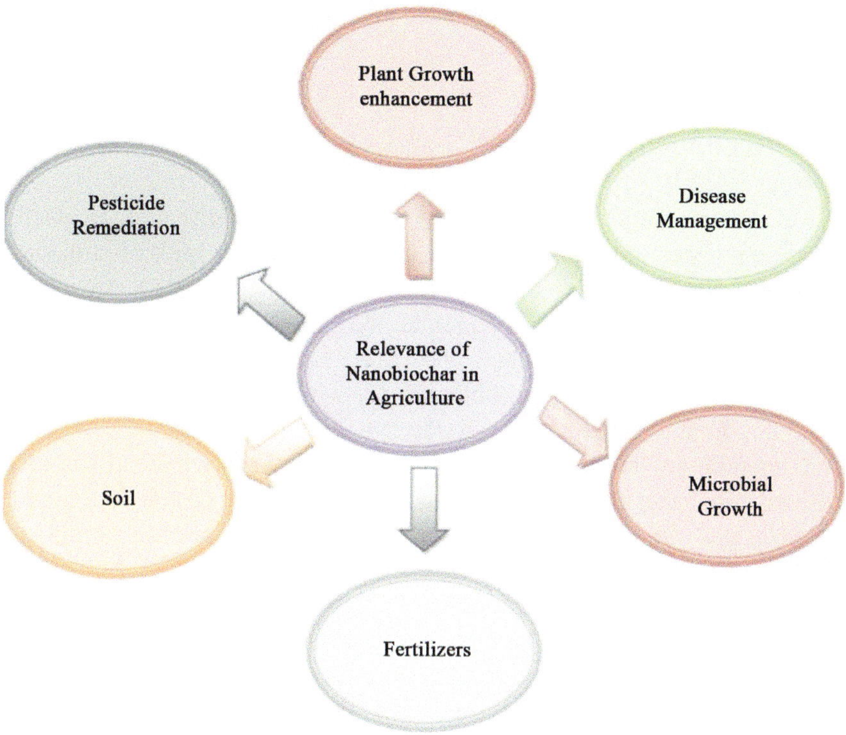

FIGURE 4.2 Significance of nano biochar composites in agriculture.

4.4 FUTURE PROSPECTS

Despite their useful applications, more research is needed to determine the extent to which nano-biochar and nanocarbon materials are hazardous to plants, animals, and the environment. Nano-biochar feedstock material, particle size, the diversity of the PAH profiles, concentration fluctuations with pyrolysis temperature, and the amount of trace metals all have an effect that must be accounted for and analyzed. More study is required to fully understand how biochar's nano-additive properties influence plant growth, microbe populations, and the natural world. Therefore, future studies will center on discovering methods to increase nano-biochar yield, electrode supercapacitor production, agglomeration reduction, and medical applications. Since these materials can be employed as nanocarriers for the delivery of pesticides, nucleic acids, and fertilizers to crops and plants, plant nanoionics is another area to investigate the potential of biochar [98–114].

4.5 CONCLUSION

Nano-biochar and biochar nanocomposites have been proven to be useful in solving agricultural and environmental issues in the bulk of published literature. Nano-biochar is an excellent adsorbent for a wide variety of contaminants, including heavy metals, dyes, pharmaceuticals, and other inorganic and organic pollutants. The Biochar Effect describes the widespread benefits of using nano-biochar for a wide range of applications, including nutrient recycling, pesticide remediation, fertilizer applications, soil amendment, and the elimination of the pathogenic effect. It has been demonstrated that the biochar effect is the consequence of the dynamic between the soil, plants, and biochar, and it ultimately causes alterations in the composition and activity of soil microorganisms. Many questions remain unanswered, yet it has already found applications in medicine, electrode objects, supercapacitors, and sensors.

KEYWORDS

- **agriculture**
- **ball-milled magnetic nano biochar**
- **cation exchange capacity**
- **electrical conductivity**
- **nano biochar**
- **oxytetracycline**
- **physicochemical properties**
- **reactive organic species**
- **soil organic carbon**
- **soil productivity**

REFERENCES

1. Atkinson, C. J., Fitzgerald, J. D., & Hipps, N. A. (2010). Potential mechanisms for achieving agricultural benefits biochar application to temperate soils: A review. *Plant and Soil, 337*(1–2), 1–18.
2. Hua, L., Lu, Z., Ma, H., & Jin, S. (2014). Effect of biochar on carbon dioxide release, organic carbon accumulation, aggregation of soil. *Environmental Progress & Sustainable Energy, 33*(3), 941–946.
3. Peterson, S. C., Jackson, M. A., & Appell, M. (2013). Biochar: Sustainable Versat. *In ACS Symposium Series, 1143*, 193–205.
4. Ulyett, J., Sakrabani, R., Kibblewhite, M., & Hann, M. (2014). Impact of biochar addition on water retention, nitrification carbon dioxide evolution two sandy loam soils. *European Journal of Soil Science, 65*(1), 96–104.
5. Lehmann, J., Rillig, M. C., Thies, J., Masiello, C. A., Hockaday, W. C., & Crowley, D. (2011). Biochar effects on soil biota – A review. *Soil Biology and Biochemistry, 43*(9), 1812–1836.
6. Chen, J., Wang, Y. Y., Wu, J. H., Si, H. P., & Lin, K. Y. (2013). The Research of Biochar Adsorption on Soil. *Applied Mechanics and Materials, 448–453*, 417–424.
7. Ahmad, M., Lee, S. S., Lim, J. E., Lee, S. E., Cho, J. S., Moon, D. H., Hashimoto, Y., & Ok, Y. S. (2014). Speciation and Phyto availability of lead and antimony in a small arms range soil amended with mussel shell, cow bone, and biochar: EXAFS spectroscopy and chemical extractions. *Chemosphere, 95*, 433–441.
8. Bai, S. H., Xu, Z., Blumfield, T. J., & Reverchon, F. (2015). Human footprints in urban forests: Implication of nitrogen deposition for nitrogen and carbon storage. *Journal of Soils and Sediments, 15*, 1927–1936.
9. Brassard, P., Godbout, S., & Raghavan, V. (2016). Soil biochar amendment as a climate change mitigation tool: Key parameters and mechanisms involved. *Journal of Environmental Management, 181*, 484–497.
10. Burrell, L. D., Zehetner, F., Rampazzo, N., Wimmer, B., & Soja, G. (2016). Long-term effects of biochar on soil physical properties. *Geoderma, 282*, 96–102.
11. Bussotti, F., Pollastrini, M., Holland, V., & Brüggemann, W. (2015). Functional traits and adaptive capacity of European forests to climate change. *Environmental and Experimental Botany, 111*, 91–113.
12. Chang, L., Wang, B. F., Liu, X. H., Callaham, M. A., & Ge, F. (2017). Recovery of collembola in *Pinus tabulaeformis* plantations. *Pedosphere, 27*, 129–137.
13. Chen, J. H., Li, S. H., Liang, C. F., Xu, Q. F., Li, Y. C., Qin, H., & Fuhrmann, J. J. (2017). Response of microbial community structure and function to short-term biochar amendment in an intensively managed bamboo (*Phyllostachys praecox*) plantation soil: Effect of particle size and addition rate. *Science of The Total Environment, 574*, 24–33.
14. Dai, Z. M., Zhang, X. J., Tang, C., Muhammad, N., Wu, J. J., Brookes, P. C., & Xu, J. M. (2017). Potential role of biochars in decreasing soil acidification—A critical review. *Science of The Total Environment, 581–582*, 601–611.
15. Deng, W., Van Zwieten, L., Lin, Z., Liu, X., Sarmah, A. K., & Wang, H. (2017). Sugarcane bagasse biochars impact respiration and greenhouse gas emissions from a latosol. *Journal of Soils and Sediments, 17*, 632–640.

16. Ding, Y., Liu, Y. G., Liu, S. B., Li, Z. W., Tan, X. F., Huang, X. X., Zheng, G. M., Zhou, L., & Zheng, B. H. (2016). Biochar to improve soil fertility: A review. *Agronomy for Sustainable Development, 36*, 36.
17. Dutta, T., Kwon, E., Bhattacharya, S. S., Jeon, B. H., Deep, A., Uchimiya, M., & Kim, K. H. (2017). Polycyclic aromatic hydrocarbons and volatile organic compounds in biochar and biochar-amended soil: A review. *GCB Bioenergy, 9*, 990–1004.
18. Lu, K., Yang, X., Shen, J., Robinson, B., Huang, H., Liu, D., Bolan, N., Pei, J., & Wang, H. (2014). Effect of bamboo and rice straw biochars on the bioavailability of Cd, Cu, Pb, and Zn to *Sedum plumbizincicola*. *Agriculture, Ecosystems & Environment, 191*, 124–132.
19. Lu, S. G., Sun, F. F., & Zong, Y. T. (2014). Effect of rice husk biochar and coal fly ash on some physical properties of expansive clayey soil (Vertisol). *Catena, 114*, 37–44.
20. Lu, K., Yang, X., Gielen, G., Bolan, N., Ok, Y. S., Niazi, N. K., Xu, S., Yuan, G., Chen, X., Zhang, X., Liu, D., Song, Z., Liu, X., & Wang, H. (2017). Effect of bamboo and rice straw biochars on the mobility and redistribution of heavy metals (Cd, Cu, Pb, and Zn) in contaminated soil. *Journal of Environmental Management, 186*(Part 2), 285–292.
21. Luo, Y., Lin, Q., Durenkamp, M., Dungait, A. J., & Brookes, P. C. (2017a). Soil priming effects following substrates addition to biochar-treated soils after 431 days of pre-incubation. *Biology and Fertility of Soils, 53*, 315–326.
22. Luo, Y., Zang, H. D., Yu, Z. Y., Chen, Z. Y., Gunina, A., Kuzyakov, Y., Xu, J. M., Zhang, K. L., & Brookes, P. C. (2017b). Priming effects in biochar enriched soils using a three-source-partitioning approach: 14C labeling and 13C natural abundance. *Soil Biology and Biochemistry, 106*, 28–35.
23. Mertens, J., Germer, J., de Araújo Filho, J. C., & Sauerborn, J. (2017). Effect of biochar, clay substrate, and manure application on water availability and tree-seedling performance in sandy soil. *Archives of Agronomy and Soil Science, 63*, 969–983.
24. Mitchell, P. J., Simpson, A. J., Soong, R., Schurman, J. S., Thomas, S. C., & Simpson, M. J. (2016). Biochar amendment and phosphorus fertilization altered forest soil microbial community and native soil organic matter molecular composition. *Biogeochemistry, 130*, 227–245.
25. Nguyen, T. T., Xu, C. Y., Tahmasbian, I., Che, R. X., Xu, Z. H., Zhou, X. H., Wallace, H. M., & Bai, S. H. (2017). Effects of biochar on soil available inorganic nitrogen: A review and meta-analysis. *Geoderma, 288*, 79–96.
26. Niazi, N. K., Bibi, I., Shahid, M., Ok, Y. S., Burton, E. D., Wang, H., Shaheen, S. M., Rinklebe, J., & Lüttge, A. (2018). Arsenic removal by perilla leaf biochar in aqueous solutions and groundwater: An integrated spectroscopic and microscopic examination. *Environmental Pollution, 232*, 31–41.
27. Qi, F., Kuppusamy, S., Naidu, R., Bolan, N. S., Ok, Y. S., Lamb, D., Li, Y., Yu, L., Semple, K. T., & Wang, H. (2017). Pyrogenic carbon and its role in contaminant immobilization in soils. *Critical Reviews in Environmental Science and Technology, 47, 795–876.*
28. Rhoades, C. C., Minatre, K. L., Pierson, D. N., Fegel, T. S., Cotrufo, M. F., & Kelly, E. F. (2017). Examining the potential of forest residue-based amendments for post-wildfire rehabilitation in Colorado, USA. *Scientifica, 2017*, Article ID 2604786.

29. Wang, Z. Y., Chen, L., Sun, F. L., Luo, X. X., Wang, H. F., Liu, G. C., Xu, Z. H., Jiang, Z. X., Pan, B., & Zheng, H. (2017). Effects of adding biochar on the properties and nitrogen bioavailability of an acidic soil. *European Journal of Soil Science, 68*, 559–572.
30. Wrobel-Tobiszewska, A., Boersma, M., Adams, P., Singh, B., Franks, S., & Sargison, J. E. (2016). Biochar for eucalyptus forestry plantations. *Acta Horticulture, 1108*, 55–62.
31. Xu, Y., Seshadri, B., Sarkar, B., Wang, H., Rumpel, C., Sparks, D., Farrell, M., Hall, T., Yang, X., & Bolan, N. (2018). Biochar modulates heavy metal toxicity and improves microbial carbon use efficiency in soil. *Science of The Total Environment, 621*, 148–159.
32. Yuan, Y., Bolan, N., Prévoteau, A., Vithanage, M., Biswas, J. K., Ok, Y. S., & Wang, H. (2017). Applications of biochar in redox-mediated reactions. *Bioresource Technology, 246*, 271–281.
33. Zhang, K., Chen, L., Li, Y., Brookes, P. C., Xu, J. M., & Luo, Y. (2016a). The effects of combinations of biochar, lime, and organic fertilizer on nitrification and nitrifiers. *Biology and Fertility of Soils, 53*, 77–87.
34. Zhang, X., Sarmah, A. K., Bolan, N. S., He, L., Lin, X., Che, L., Tang, C., & Wang, H. (2016b). Effect of the aging process on adsorption of diethyl phthalate in soils amended with bamboo biochar. *Chemosphere, 142*, 28–34.
35. Shaaban, M., Van Zwieten, L., Bashir, S., Younas, A., Núñez-Delgado, A., Chhajro, M. A., Kubar, K. A., Ali, U., Rana, M. S., Hu, R., (2018). A concise review of biochar application to agricultural soils to improve soil conditions and fight pollution. *Journal of Environmental Management, 28*, 429–440.
36. Li, Y., Hu, S., Chen, J., et al. (2018). Effects of biochar application in forest ecosystems on soil properties and greenhouse gas emissions: A review. *Journal of Soils and Sediments, 18*, 546–563.
37. Lehmann, J. (2007). A handful of carbon. *Nature, 447*, 143–144.
38. Alimohammadi, M., Xu, Y., Wang, D., Biris, A. S., & Khodakovskaya, M. V. (2011). Physiological responses induced in tomato plants by a two-component nano structural system composed of carbon nanotubes conjugated with quantum dots and its in vivo multimodal detection. *Nanotechnology, 22*(29), 295101.
39. Tripathi, S., Sonkar, S. K., & Sarkar, S. (2011). Growth stimulation of gram (*Cicer arietinum*) plant by water-soluble carbon nanotubes. *Nanoscale, 3*, 1176–1181.
40. Ashfaq, M., Verma, N., & Khan, S. (2017). Carbon nanofibers as a micronutrient carrier in plants: Efficient translocation and controlled release of Cu nanoparticles. *Environmental Science: Nano, 4*, 138–148.
41. Khodakovskaya, M. V., Kim, B. S., Kim, J. N., et al. (2013). Carbon nanotubes as plant growth regulators: Effects on tomato growth, reproductive system, and soil microbial community. *Small, 9*, 115–123.
42. Lehmann, J., Gaunt, J., & Rondon, M. (2006). Biochar sequestration in terrestrial ecosystems—A review. *Mitigation and Adaptation Strategies for Global Change, 11*, 403–427.
43. Sohi, S. P. (2012). Carbon storage with benefits. *Science, 338*, 1034–1035.
44. Saxena, J., Rawat, J., & Kumar, R. (2017). Conversion of biomass waste into biochar and the effect on mung bean crop production. *Clean, 45*, 1501020.

45. Ashiq, A., Adassooriya, N. M., Sarkar, B., Rajapaksha, A. U., Ok, Y. S., & Vithanage, M. (2019). Municipal solid waste biochar-bentonite composite for the removal of antibiotic ciprofloxacin from aqueous media. *Journal of Environmental Management, 236*, 428–435.
46. Panwar, N. L., Pawar, A., & Salvi, B. L. (2019). Comprehensive review on production and utilization of biochar. *SN Applied Sciences, 1*, 168.
47. Ashiq, A., & Vithanage, M. (2020). Biochar-mediated soils for efficient use of agrochemicals. In M. Narasimha & V. Prasad (Eds.), *Agrochemicals Detection, Treatment and Remediation, Pesticides and Chemical Fertilizers* (pp. 621–645). Elsevier BV.
48. Noreen, S., & Abd-Elsalam, K. A. (2021). Biochar-based nanocomposites: A sustainable tool in wastewater bioremediation. In K. A. Abd-Elsalam & M. Zahid (Eds.), *Applications of Nanomaterials for Water Purification, Micro and Nano Technologies, Aqua nanotechnology* (pp. 185–200). Elsevier BV.
49. Ramanayaka, S., Vithanage, M., Alessi, D., Liu, W. J., Jayasundera, A. C. A., & Ok, Y. S. (2010). Nano biochar: Production, properties, and multifunctional applications. *Environmental Science: Nano, 7*, 3279–3302.
50. Wang, D., Zhang, W., Hao, X., & Zhou, D. (2013). Transport of biochar particles in saturated granular media: Effects of pyrolysis temperature and particle size. *Environmental Science & Technology, 47*, 821–828.
51. Gamiz, B., Cox, L., Hermosin, M. C., Spokas, K., & Celis, R. (2017). Assessing the effect of organoclays and biochar on the fate of abscisic acid in soil. *Journal of Agricultural and Food Chemistry, 65*, 29–38.
52. Tan, X. F., Liu, Y. G., Zeng, G., et al. (2015). Application of biochar for the removal of pollutants from aqueous solutions. *Chemosphere, 125*, 70–85.
53. Tan, X. F., Liu, Y. G., Gu, Y. L., et al. (2016). Biochar-based nano-composites for the decontamination of wastewater: A review. *Bioresource Technology, 212*, 318–333.
54. Yao, Y., Gao, B., Chen, J., & Yang, L. (2013). Engineered biochar reclaiming phosphate from aqueous solutions: Mechanisms and potential application as a slow-release fertilizer. *Environmental Science & Technology, 47*, 8700–8708.
55. Yao, Y., Gao, V., Inyang, M., et al. (2011). Biochar derived from anaerobically digested sugar beet tailings: Characterization and phosphate removal potential. *Bioresource Technology, 102*, 6273–6278.
56. Zhang, M., Gao, B., Varnoosfaderani, S., Hebard, A., Yao, Y., & Inyang, M. (2013). Preparation and characterization of a novel magnetic biochar for arsenic removal. *Bioresource Technology, 130*, 457–462.
57. Inyang, M., Gao, B., Zimmerman, A., Zhang, M., & Chen, H. (2014). Synthesis, characterization, and dye sorption ability of carbon nanotube–biochar nanocomposites. *Chemical Engineering Journal, 236*, 39–46.
58. Naghdi, M., Taheran, M., Brar, S. K., et al. (2018). Pinewood nano biochar: A unique carrier for the immobilization of crude laccase by covalent bonding. *International Journal of Biological Macromolecules, 115*, 563–571.
59. Ramadan, M. M., Amal, A., & Abd-Elsalam, K. A. (2020). Micro/nano biochar for sustainable plant health: Present status and prospects. In A. Kamel & K. A. Abd-Elsalam (Eds.), *Carbon Nanomaterials for Agri-Food and Environmental Applications* (pp. 323–357). Elsevier BV.

60. Ehsan, K. (2020). Enhanced light olefins production via methanol dehydration over promoted SAPO-34. In *Advances in Chemistry Research, Volume 63, Chapter: 4.* Nova Science Publishers, Inc., NY, USA.
61. Ehsan, K. (2020). Gas hydrate: Applications, structure, formation, separation processes, thermodynamics. In J. C. Taylor (Ed.), *Advances in Chemistry Research, Volume 62, Chapter: 8* (pp. 1–21). Nova Science Publishers, Inc., NY, USA.
62. Mehran, K., Farshid, K., & Ehsan, K. (2016). The Effect of Nano-Composites on the Mechanic and Morphological Characteristics of NBR/PA6 Blends. *American Journal of Oil and Chemical Technology, 4*(1), 29–44.
63. Ehsan, K. (2016). The Effect of Nano-Composites on the Mechanic and Morphological Characteristics of NBR/PA6 Blends. *American Journal of Oil and Chemical Technology, 4*(1), 27–42.
64. Farshad Kianfa, Seyed Reza, Mahdavi Moghadam, & Ehsan Kianfar. (2015). Energy optimization of Ilam Gas Refinery Unit 100 by using HYSYS Refinery Software. *Indian Journal of Science and Technology, 8*(S9), 431–436.
65. Li, Y. C., Li, Y. F., Chang, S. X., Liang, X., Qin, H., Chen, J. H., & Xu, Q. F. (2017). Linking soil fungal community structure and function to soil organic carbon chemical composition in intensively managed subtropical bamboo forests. *Soil Biology and Biochemistry, 107*, 19–31.
66. Li, Z. G., Gu, C. M., Zhang, R. H., Ibrahim, M., Zhang, G. S., Wang, L., & Liu, Y. (2017). The beneficial effect induced by biochar on soil erosion and nutrient loss of sloping land under natural rainfall conditions in central China. *Agricultural Water Management, 185*, 145–150.
67. Lin, Z. B., Liu, Q., Liu, G., Cowie, A. L., Bei, Q. C., Liu, B. J., & Xie, Z. B. (2017). Effects of different biochars on *Pinus elliottii* growth, N use efficiency, soil N_2O and CH4 emissions and C storage in a subtropical area of China. *Pedosphere, 27*, 248–261.
68. Ma, S. Q., Jing, F. Q., Sohi, S. P., & Chen, J. W. (2019). New insights into contrasting mechanisms for PAE adsorption on millimeter, micron, and nano-scale biochar. *Environmental Science and Pollution Research, 26*, 18636–18650.
69. Samsudin, M. H., Hassan, M. A., Idris, J., et al. (2019). A one-step self-sustained low-temperature carbonization of coconut shell biomass produced a high specific surface area biochar-derived nano-adsorbent. *Waste Management Research, 37*, 551–555.
70. Ramanayaka, S., Tsang, D. C. W., Hou, D., et al. (2019). Green synthesis of graphitic nano biochar for the removal of emerging contaminants in aqueous media. *Science of The Total Environment, 310*, 107306.
71. Li, C., Zhang, L., Gao, Y., & Li, A. (2018). Facile synthesis of nano ZnO/ZnS modified biochar by directly pyrolyzing zinc-contaminated corn stover for Pb (II), Cu (II), and Cr (VI) removals. *Waste Management, 79*, 625–637.
72. Zhu, Y., Hu, J., & Wang, J. (2012). Competitive adsorption of Pb (II), Cu (II), and Zn (II) onto xanthate-modified magnetic chitosan. *Journal of Hazardous Materials, 155–161*, 221.
73. Guo, S., Dan, Z., Duan, N., et al. (2018). Zn (II), Pb (II), and Cd (II) adsorption from aqueous solution by magnetic silica gel: Preparation, characterization, and adsorption. *Environmental Science and Pollution Research, 25*, 30938–30948.

74. Gan, C., Liu, Y. G., Tan, X. F., et al. (2015). Effect of porous zinc–Biochar nanocomposites on Cr (VI) adsorption from aqueous solution. *RSC Advances, 5*, 35107–35115.
75. Li, R., Zhang, Y., Deng, H., Zhang, Z., Wang, J. J., Shaheen, S. M., & Du, J. (2019). Removing tetracycline and Hg (II) with ball-milled magnetic nano biochar and its potential on polluted irrigation water reclamation. *Journal of Hazardous Materials, 384, 121095.*
76. He, S., Zhong, L., Duan, J., Feng, Y., Yang, B., & Yang, L. (2017). Bioremediation of wastewater by iron oxide-biochar nanocomposites loaded with photosynthetic bacteria. *Frontiers in Microbiology, 8*, 823.
77. Kianfar, E. (2020). Magnetic nanoparticles in targeted drug delivery: A review. *Journal of Superconductivity and Novel Magnetism, 34, 1709–1735.*
78. Syah, Rahmad, Zahar, Marziah & Kianfar, Ehsan. (2021). "Nanoreactors: properties, applications and characterization" *International Journal of Chemical Reactor Engineering, 19*(10), 981–1007. https://doi.org/10.1515/ijcre-2021-0069.
79. Majdi, H. S., Latipov, Z. A., Borisov, V., et al. (2021). Nano and Battery Anode: A Review. *Nanoscale Research Letters, 16*, 177.
80. Dmitry, B., Abduladheem, T. J., Supat, C., Wanich, S., Mohammad, J. A., Iman, H., & Valiev, E. K. (2021). Nanomaterial by sol-gel method: Synthesis and application. *Advances in Materials Science and Engineering, 5102014*, 1–21.
81. Ansari, M. J., Kadhim, M. M., Hussein, B. A., et al. (2022). Synthesis and stability of magnetic nanoparticles. *Bio Nano Science, 12, 627–638.*
82. Supat, C., Kavitha, M., Wanich, S., Mohammad, J. A., Zuhair, I., Al, M., Mustafa, M. K., Yasser, F. M., Shafik, S., & Shafik, E. K. (2022). Morphological control: Properties and applications of metal nanostructures. *Advances in Materials Science and Engineering, 1971891*, 1–15.
83. Omer, D. A., Salah, A., Mustafa, Z. M., Hasan, S. M., Dhameer, A. M., Khusniddin, F. U., & Ehsan, K. (2022). Investigation of effective parameters Ce and Zr in the synthesis of H-ZSM-5 and SAPO-34 on the production of light olefins from naphtha. *Advances in Materials Science and Engineering, 6165180*, 1–22.
84. Asep, S., Indah, R., Lakshmi, T., Firas, R. A., Mustafa, M. K., Usama, S., Altimari, Z. H. M., Yasser, F. M., & Ehsan, K. (2022). A review of high-energy density lithium-air battery technology: Investigating the effect of oxides and nanocatalysts. *Journal of Chemistry, 2762647*, 1–32.
85. Abdelbasset, W. K., Jasim, S. A., Bokov, D. O., et al. (2022). Comparison and evaluation of the performance of graphene-based biosensors. *Carbon Letters, 32, 927–951.*
86. Jasim, S. A., Kadhim, M. M., KN, V., et al. (2022). Molecular junctions: Introduction and physical foundations, nanoelectrical conductivity and electronic structure and charge transfer in organic molecular junctions. *Brazilian Journal of Physics, 52*, 31.
87. Dehghan, M., Tahmasebipour, M., & Ebrahimi, S. (2022). Design, fabrication, and characterization of an SLA 3D printed nanocomposite electromagnetic microactuator. *Microelectronic Engineering, 254*, 111695.
88. Shakir A. R., Muhsin K. H., & Almsaid, H. (2022). Rat hepatocellular primary cells: A cellular and genetic assessment of the chitosan nanoparticles-induced damage and

cytotoxicity. *Archives of Razi Institute, 77*(2), 579–584. https://doi.org/10.22092/ari.2022.357103.1974.

89. Issa, M. (2022). Rapid enzymatically reduction of Zincum, Gluconicum for the biomanufacturing of zinc oxide nanoparticles by Mycoextracellular filtrate of Penicillium Digitatum (Pdig-B3) as a soft green technique. *Archives of Razi Institute, 77*(1), 101–110. https://doi.org/10.22092/ari.2021.356422.1841.
90. Fakri, M. Y., Riyadh, K. R., Tareq, M. E., Bashir, M., & Khudhayer, O. M. (2021). Effects of structural manipulation on the bioactivity of some coumarin-based products. *Archives of Razi Institute, 76*(5), 1297–1305. https://doi.org/10.22092/ari.2021.356100.1776.
91. Mohammad, M., Ahmed, S., & Alameri, R. (2021). Green medicine: A novel preparation method for green synthesizing of iron nanoparticles derived from *Beta vulgaris* extract. *Archives of Razi Institute, 76*(5), 1327–1332. https://doi.org/10.22092/ari.2021.355933.1740.
92. Saied, S., Mohammed, S., Khaleel, B., & Saleh, M. (2021). Comparative studies between conventional techniques and green chemistry to the synthesis of novel piperidinium salts ionic liquids (PBSILs). *Journal of Chemical Health Risks, 11*(4), 451–456. https://doi.org/10.22034/jchr.2021.686640.
93. Kundu, R., Biswas, C., Ahmed, J., Naime, J., & Ara, M. (2020). A study on the adsorption of cadmium (II) from aqueous solution onto activated carbon originated from *Bombax ceiba* fruit shell. *Journal of Chemical Health Risks, 10*(4), 243–252. https://doi.org/10.22034/jchr.2020.1903764.1154.
94. Honarvar, Z., Farhoodi, M., Khani, M., Mohammadi, A., Shokri, B., Jannatiha, N., & Shojaee-Aliabadi, S. (2021). Antimicrobial and physicochemical properties of plasma treated bio-coating polypropylene films containing *Satureja hortensis* essential oil. *Iranian Journal of Chemistry and Chemical Engineering, 40*(4), 1216–1228. https://doi.org/10.30492/ijcce.2020.43324.
95. Aramesh, B. Z., & Asadi, A. Z. (2021). Electrochemical determination venlafaxine at NiO/GR nanocomposite modified carbon paste electrode. *Iranian Journal of Chemistry and Chemical Engineering, 40*(4), 1042–1053. https://doi.org/10.30492/ijcce.2020.39188.
96. Musa, A. A., Salim, F. B., Shiaka, G. P., Muhammad, M., Nafisa, B., Surayya, M. M., Sa'adatu, A. Y., Hauwa, S. D., Gayus, R., & Gumel, A. M. (2021). Influence of reaction temperature on bioethanol production by *Saccharomyces cerevisiae* using cassava as substrate. *International Journal of Sustainable Energy and Environmental Research, 10*(1), 9–16.
97. Rajput, V. D., Minkina, T., Kumari, A., Harish, Singh, V. K., Verma, K. K., Mandzhieva, S., Sushkova, S., Srivastava, S., & Keswani, C. (2021). Coping with the challenges of abiotic stress in plants: New dimensions in the field application of nanoparticles. *Plants, 10*(6), 1221. https://doi.org/10.3390/plants10061221.
98. Rajput, V., Minkina, T., Sushkova, S., Behal, A., Maksimov, A., Blicharska, E., Ghazaryan, K., Movsesyan, H., & Barsova, N. (2019). ZnO and CuO nanoparticles: A threat to soil organisms, plants, and human health. *Environmental Geochemistry and Health, 42*(1), 147–158. https://doi.org/10.1007/s10653-019-00317-3.
99. Rajput, V. D., Minkina, T., Sushkova, S., Tsitsuashvili, V., Mandzhieva, S., Gorovtsov, A., Nevidomskyaya, D., & Gromakova, N. (2017). Effect of nanoparticles on crops

and soil microbial communities. *Journal of Soils and Sediments, 18*(6), 2179–2187. https://doi.org/10.1007/s11368-017-1793-2.

100. Rajput, V., Minkina, T., Fedorenko, A., Sushkova, S., Mandzhieva, S., Lysenko, V., Duplii, N., Fedorenko, G., Dvadnenko, K., & Ghazaryan, K. (2018). Toxicity of copper oxide nanoparticles on spring barley (*Hordeum sativum* distichum). *Science of The Total Environment, 645*, 1103–1113. https://doi.org/10.1016/j.scitotenv.2018.07.211.
101. Rajput, V. D., Kumari, A., Minkina, T., Barakhov, A., Singh, S., Mandzhieva, S. S., Sushkova, S., Ranjan, A., Rajput, P., & Garg, M. C. (2022). A practical evaluation of the integrated role of biochar and nanomaterials in soil remediation processes. *Environmental Geochemistry and Health, 45,* 9435–9449. https://doi.org/10.1007/s10653-022-01375-w.
102. Rajput, V. D., Faizan, M., Upadhyay, S. K., Kumari, A., Ranjan, A., Sushkova, S., Chauhan, P. K., Mahmud, A. A., Burachevskaya, M., Chaplygin, V., & Deryabkina, I. (2022). Influence of nanoparticles on the plant rhizosphere microbiome. In T. Minkina (Ed.), *Sustainable Plant Nutrition in a Changing World* (pp. 83–102). https://doi.org/10.1007/978-3-030-97389-6_4.
103. Rajput, V. D., Singh, A., Minkina, T., Rawat, S., Mandzhieva, S., Sushkova, S., Shuvaeva, V., Nazarenko, O., Rajput, P., Komariah, & Verma, K. K., et al. (2021). Nano-enabled products: Challenges and opportunities for sustainable agriculture. *Plants, 10*(12), 2727. https://doi.org/10.3390/plants10122727.
104. Singh, A., Singh, A. K., Rawat, S., Pal, N., Rajput, V. D., Minkina, T., Sharma, R., Singh, N. P., & Tripathi, J. N. (2022). Satellite-based quantification of methane emissions from wetlands and rice paddies ecosystems in north and northeast India. *Hydrobiologia, 1*(3), 317–330. https://doi.org/10.3390/hydrobiology1030023.
105. Singh, A., Sengar, R. S., Rajput, V. D., Minkina, T., & Singh, R. K. (2022). Zinc oxide nanoparticles improve salt tolerance in rice seedlings by improving physiological and biochemical indices. *Agriculture, 12*(7), 1014. https://doi.org/10.3390/agriculture12071014.
106. Singh, A., Sengar, R. S., Sharma, R., & Rajput, P. (2021). Nano-priming technology for sustainable agriculture. *Biogeosystem Technology, 8*(2), 79–92. https://doi.org/10.13187/bgt.2021.2.79.
107. Singh, A., Mehrotra, R., Rajput, V. D., Dmitriev, P., Singh, A. K., Kumar, P., Tomar, R. S., Singh, O., & Singh, A. K. (2022). Geoinformatics, artificial intelligence, sensor technology, big data. In D. Deepika (Ed.), *Sustainable Agriculture Systems and Technology* (pp. 295–313). https://doi.org/10.1002/9781119808565.ch14.
108. Singh, A., Rajput, V. D., Rawat, S., Sharma, R., Singh, A. K., Kumar, P., Singh, A. K., Minkina, T., Singh, R. P., & Singh, S. (2022). Geoinformatics and nanotechnological approaches for coping up abiotic and biotic stress in crop plants. In D. Deepika (Ed.), *Sustainable Agriculture Systems and Technology* (pp. 337–359). https://doi.org/10.1002/9781119808565.ch17.
109. Singh, A., Rajput, V. D., Rawat, S., Sharma, R. K., Singh, A. K., & Singh, A. S. (2021). Emerging tools for sustainable agriculture and food security: An introduction. *Emerging Tools for Sustainable Agriculture and Food Security, 1*(1), 3–16.
110. Rajput, P., Rajput, V. D., Singh, A., Sharma, R., Rawat, S. S., Tomar, R., Singh, A., & Minkina, T. (2021). Impact of emerging tools for solving 'hidden hunger' in humans. *Emerging Tools for Sustainable Agriculture and Food Security, 1*(1), 17–30.

111. Rajput, V. D., Minkina, T., Kumari, A., Shende, S. S., Ranjan, A., Faizan, M., Barakvov, A., Gromovik, A., Gorbunova, N., Rajput, P., et al. (2022). A review on nanobioremediation approaches for restoration of contaminated soil. *Eurasian Journal of Soil Science, 11*(1), 43–60. https://doi.org/10.18393/ejss.990605.
112. Rajput, V. D., Minkina, T., Feizi, M., Kumari, A., Khan, M., Mandzhieva, S., Sushkova, S., El-Ramady, H., Verma, K. K., Singh, A., et al. (2021). Effects of silicon and silicon-based nanoparticles on rhizosphere microbiome, plant stress and growth. *Biology, 10*(8), 791. https://doi.org/10.3390/biology10080791.
113. Singh, A., et al. (2020). Monitoring soil salinity and recent advances in the mechanism of salinity tolerance in plants. *Biogeosystem Technique, 7*(2), 66–87. https://doi.org/10.13187/bgt.2020.2.66.
114. Singh, A., Sengar, R. S., Shahi, U. P., Rajput, V. D., Minkina, T., & Ghazaryan, K. A. (2022). Prominent effects of zinc oxide nanoparticles on roots of rice (*Oryza sativa* L.) grown under salinity stress. *Stresses, 3*, 33–46. MDPI AG. https://doi.org/10.3390/stresses3010004.

CHAPTER 5

Role of Nanoparticles and Nano-Biochar in Reducing Salinity Stress

NIHAD ASHRAF KHAN

Department of Biosciences, Jamia Millia Islamia University, New Delhi, India

ABSTRACT

Management of crop residue has always been a challenging task for farmers to maintain a balance between the economy and a sustainable environment. The continuous practice of burning crop residue has contributed to an increase in greenhouse gases in the atmosphere and the depletion of nutrients from the soil. Excessive concentration of salts in the soil reporting electrical conductivity (EC) of more than 20mM affects the growth of plants and leads to osmotic stress and loss of vegetation in arid and semi-arid regions. The main driving forces for naturally-induced soil salinity are the intrusion of highly salinized water in coastal areas, aeolian, and dissolution of soil parent minerals. With the advent of nanotechnology, Nano-Biochar produced from specific raw materials by the process of pyrolysis operated at different temperatures has proven to be an emerging sorbent for sustainable agriculture and pollutant remediation by acting as a soil ameliorant and crop enhancer. The nanostructure of the NPs and nano-biochar helps to reduce evapotranspiration and increases aeration and soil water-holding capacity. This results in the dilution of salt ions concentration under salinity stress and thus maintains a suitable soil environment for plant growth. It helps in the adsorption of excess Na_+ in soil and increases in mineral content of soil by releasing K_+, Ca_{2+}, and Mg_{2+} which even interfere with the uptake of Na_+ by plants. However, the

Nano-Biochar and Nanoparticles for Soil Health Improvement.
Vishnu D. Rajput and Tatiana M. Minkina (Eds.)

long-term effect of the NPs and nano-biochar and their relation with the biogeochemical cycle needs to be evaluated to bring out the large-scale application of these components for future use.

5.1 INTRODUCTION

A sustainable approach in the agricultural sector through biomass conversion has been widely explored for the production of sustainable materials such as biochar (BC), activated carbon, graphene, etc. [1]. BC is a carbonaceous source produced from the thermochemical conversion of biomasses of plants and animals operated under high temperatures, in the absence of or limited oxygen availability. BC has been employed as a soil supplement globally and has shown importance in the areas of agro-industries, agri-sectors, and environmental remediation, providing a solution to the agriculture and environmental protection sectors. With the advent of nanotechnology focus on the production of nano-sized BC (less than 100 μm) has become more significant due to its various physicochemical attributes of great importance for sustainable agriculture [2]. Nano-BC has a large surface area due to its small size, high porosity, surface hydrophobicity, aromaticity/polarity, and zeta potential. These characteristics of Nano-BC make the adsorption rate highly efficient and even help to neutralize low pH soils due to their alkaline nature [2]. Due to the increased salinity stress of soil in coastal areas, a need for a sustainable approach has found its way through Nano-BCs. The ongoing climate changes due to anthropogenic activities and the excess use of fertilizers have led to soil stress and negative implications on the environment [3]. The abiotic soil stress due to increased concentration of salt in water resources (fertilizers, industrial waste, pesticides) or soil extracts overexposure of one to two weeks decreases soil fertility, absorption efficiency, and high yield production. Thus, impacting the health of a plant and prominently leading to plant injury [4]. The nature of a BC depends mostly on the kind of feedstock materials used for its production and pyrolysis conditions which markedly change the morphology and durability of a BC. To optimize the size of the BC, two factors play an important role: increasing pyrolysis temperature and grinding time. Additionally, nano-BC possesses large fractions of carbon, oxygen, and hydrogen forming several phenolic components [5]. Various nanoparticles such as silicon, manganese, zinc, magnesium, and carbon produced from top-down and bottom-up approaches provide a promising effect on the sustainable development of the agricultural sector in mitigating the salinity stress on plants.

Several different soil stresses in plants are caused by high salt concentration, drought conditions, over-flooding with water, and mechanical and environmental stresses. One of the rising concerns in the agricultural and environmental remediation sector is the salinization of soil and increasing demand for crop yield due to the rapid expansion of the human population. Thus, improving soil quality and yield has become a challenging task worldwide. It has been reported that approximately 20% of the world's irrigated land, exclusively used for agriculture, is affected by soil salinization [6]. The adverse implications of salt stress on plants include inhibition of seed germination, improper growth and development, and delayed flowering and fruiting [7]. The increased concentration of Na_+ ions beyond a tolerance level limits the uptake of water by plants and the absorption of nutrients in the plants [8]. Water deficiency coupled with nutritional imbalance induces two primary stresses: ionic stress and osmotic stress, which alter major physiological mechanisms in plants [9]. NPs and nano-BC prove to be efficient tools in agriculture for reducing various soil stresses impacting the health of plants and helping in improving the soil's biological characteristics and absorption of nutrients. It helps in the immobilization, absorption, and controlled production of antioxidants under trace metal stress [10]. Supplementing soil with nano-BC provides an appropriate environment for the microbes to flourish, thereby increasing the soil fertility, quality, tensile strength, porosity, and water retention capability of soil for stabilizing soil activities [11, 12]. Nano-BC has the capability of restoring degraded soils and is a useful resource in the environmental remediation sector where prominent causes of eutrophication, surface runoff, fertilizers, and industrial emissions debilitate the soil quality [13, 14].

This article focuses on the agricultural and environmentally friendly properties of NPs and nano-BC for sustainable farming. It also highlights the role of NPs and nano-BC in reducing salinity stress, which negatively affects plant health and agricultural production on a large scale.

5.2 PRODUCTION AND CHARACTERISTICS OF NANO-BIOCHAR

5.2.1 PRODUCTION

The characteristics of a Nano-BC differ from bulk-BC largely based on its extremely small size, surface properties, efficient adsorption rate, mobilization, and bioavailability. The mechanism through which a nano-sized BC can be achieved includes ball milling, disc milling, and pyrolysis at high temperatures explained in Table 5.1 [15]. Thus, the most commonly used method assisted with milling and pyrolysis is mechanical grinding. The process of mechanical grinding was reported in the production of graphite nano-BC through the conversion of woody BC derived from *Gliricidia septum* by Ramanayaka et al. [16]. During the conversion of rice husk to nano-BC by Nath et al. [17], the process of milling was performed at 500 rotations per minute under an ambient temperature in a confined container until finely ground to nano-sized particles (Figure 5.1). The nano-BC was further chemically modified for remediation purposes [17, 18].

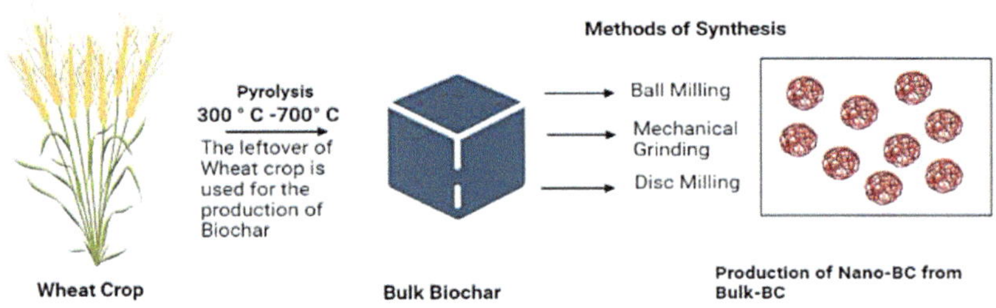

FIGURE 5.1 Representation of various steps involved in nano-BC production from a wheat crop.

5.3 CHARACTERISTICS

Nano-BC, a rich carbon-based sustainable material, has particular characteristics determined by the kind of feedstock used for its production and methods involved in its synthesis, such as pyrolysis, ball milling, grinding time, chemical modifications, etc. The different kinds of feedstocks normally used are wheat straw, pine wood, rice husk, etc. [19, 20]. Nano-BC derived from lignin-containing materials has hydroxyl, carboxylic, carbonyl, and ether functional groups present. These functional groups provide stretchability, strength, and bending properties to nano-BC [21]. Nano-BC is small-sized compared to bulk-BC and has a large surface area, EC, and X-ray diffraction patterns [19]. The main difference in the characteristics of various nano-BCs is based on functional groups present on

TABLE 5.1 Compilation of Certain Sustainable Materials Used for the Production of Nano-BC and Their Useful Environmental Impact

Sl. No.	Sustainable Materials	Methods	Product Obtained	Useful Impact	References
1.	Rice-husk	Pyrolysis at 600°C	Rice-husk nano-BC permeated with Iron oxide.	Adsorption and removal of arsenic from the aqueous media.	[17]
2.	Pinewood	Ball milling	Nano-BC (size 60 nm)	Attachment of carbamazepine of the surface for effective removal.	[19]
3.	Woody biochar	Disc milling at –80°C in ethanol media.	Graphite nano-BC	Efficient removal of cadmium (CdII), and chromium (CrIII).	[16]
4.	Wheat straw	Ball milling at 700°C.	Magnetic nano-BC.	Adsorption of Hg and tetracycline from aqueous media.	[18]
5.	Corncob bares	Pyrolysis at 400°C under low oxygen availability.	Triethylenetetramine modified nano-BC.	It helps to remove dyes and tartazine from the aqueous media.	[20]

the surface, specific surface area, and different sizes (nm) of the nano-BC measured with the help of an electron microscope. Under photoelectron spectroscopy and infrared spectroscopy, a detailed examination of the characteristics and SSA of nano-BC can be performed [22].

5.3.1 PROMINENT CAUSES OF SOIL SALINITY

Environment sustainability and food security have been greatly hindered globally due to soil salinity caused either by natural or anthropogenic activities. The developmental process of increasing soil quality and its nourishment has deteriorated due to climate changes [23]. Apart from salinization, the key elements that affect soil formation are temperature, precipitation, wind, metal toxicity, water stress, and inefficient soil-microbe interactions. Arid regions are more prone to soil salinization as a result of salt accumulation in the soil solum. Under the effect of salinity, the soil initially undergoes different stages which ultimately result in unproductive and barren land. Increased concentration of salt at an initial stage (Na_{+}, K_{+}, Cl_{-}, Ca_{2+}) affects the soil-microbe interactions, thus reducing the productivity of the soil gradually. In later stages, it drastically affects the soil vegetation and other organisms in the soil required for soil renewal and fertility [24].

5.3.2 NATURAL CAUSES

The weathering of rocks and minerals releases several soluble cations and anions such as Ca_{2+}, Na_{+}, K_{+}, Cl_{-}, HCO_{3-}, and SO_{4-} in the environment [25]. During arid and semi-arid conditions, calcium carbonate ($CaCO_3$), gypsum ($CaSO_4.2H_2O$), and magnesite ($MgCO_3$), which are usually less soluble, become more soluble. In arid regions, high evaporation rates concentrate water solution, precipitating salts in the soil. The primary cause of soil salinization is attributed to coastal areas that are found in arid and semi-arid regions. Soil salinization in oceanic islands results from the transfer of salts from the oceanic surface to the ground surface [26].

Another prominent cause of soil salinity in nature is regarded as volcanic activities during which wind-blown ashes having sodium-rich minerals result in the salinization of soil [27]. Moreover, the salinity of groundwater and soil is further increased by fumaroles and thermal springs that contain chlorine and sulfur [28]. In short, soil salinity is anticipated to rise due to the impact of global warming.

5.3.3 ANTHROPOGENIC CAUSES

With the advent of technology and advancements in farming, agricultural complexity has taken a new route. Increasing the facility of irrigating huge hectares of agricultural lands in a short period has its shortcomings. Irrigation with improper leaching and drainage has become the secondary cause of soil salinization. Accumulation of these dissolved salts in the soil reaches the ground surface due to prolonged water-logging in irrigation fields that impedes plant growth [29]. Water flowing downstream into the rivers from the irrigated fields leads to an increase in salt concentration in the river. Utilizing this river water to irrigate agricultural areas downstream hastens the salinization process of the land [30]. Soil classification and the relation of salinity level (dCm_1) with crop yield have been discussed in Table 5.2 [31].

From a record high of 915.5 million ha in 1986 to 1,069 million ha in 2016, soil salinization is rising globally. It is anticipated that by 2050, 50% of the world's croplands will experience varied degrees of salt, increasing soil salinization [32]. Decreased productivity and crop yield due to salinization reduce economic benefits and overall agricultural revenue by 7.9% [33]. Sodification of soil as a result of increased Na_+ ions deteriorates the soil structure, which not only affects the infiltration of water but also interferes with the penetration of air into the soil profile. Thus, limiting the availability of water and oxygen for plants. Moreover, decreasing the absorption of nutrients for the nourishment and proper development of plants [34]. Hence, the concern regarding salinization and sodification of soil needs to be addressed, and certain alternatives for the betterment of agricultural and environmental remediation sectors need to be programmed for a multi-fold production of crop yield in the future.

TABLE 5.2 Soil Salinity and Effect on Crop Health

Level of Salinization	Level of Salinity EC (dCm1)	Effect on Crop Yield
No-salinity	0–2	No negative effect on the crop.
Less-salinity	2–4	Crops sensitive to salinity show reduced productivity and yield.
Moderate-salinity	4–8	The yield of a large group of crops gets restricted.
Strong-salinity	8–16	Crops tolerant to salinity show reduced productivity and yield.
Highest-salinity	> 16	This range of salinity is mostly suitable for halophytes.

5.4 MECHANISM AND PHYSIOLOGY INVOLVED IN PLANTS IN RESPONSE TO SALINITY STRESS

The plant-soil relationship is the governing factor for environmental stability. Plants, being sessile, have to develop several mechanisms to tolerate the adversity of the climate. They have developed mechanisms to tolerate salinity stress, imbalance in the microbial biome of the soil, metallic stress, temperature stress, etc. One of the major causes of the impediment to plant growth is salinity stress. To flourish even in such adverse conditions, several mechanisms and physiological changes take place within plants in response to salinity stress caused by natural or anthropogenic interventions.

In nature, plants exist as halophytes and glycophytes. Plants tolerant towards salinity stress and having optimized growth levels even under high salt concentration are regarded as halophytes [35], whereas plants sensitive to salt conditions, having decreased growth and development, are therefore regarded as glycophytes. Interestingly, in nature, the majority of plants are glycophytes [36]. An increased level of Na_{+} ions in the soil impacts water and nutrition absorption of plants. This leads to osmotic and ionic stress [37], which majorly affects the light-harvesting complex formation required for photosynthesis, repression in cell division, and expansion, and alteration in sugar signaling in plants [38, 39, 40]. To mitigate the effects of high salt concentration, plants have developed several mechanisms to adapt to saline conditions. These strategies include stress sensing and expression of various salt-stress responsive genes, coupled with complex signal transduction pathways that regulate osmotic, ionic, homeostatic, and detoxification transports in plants [38, 41].

5.4.1 ROLE OF SALT-STRESS SENSING IN PLANTS

The three most important factors that trigger a salt-stress response in plants include increased Na_{+} ion concentration, excess intracellular Ca_{2+} levels, and formation of excess reactive oxygen species (ROS) [43]. This salt stress leads to osmotic stress and activation of Ca_{2+} channels where Ca_{2+} binds with the receptor, acts as a second messenger, and evokes a cascade of calcium signal response [44]. Calcium-permeable channel OSCA1 [45] and K_{+} exchange anti-porters KEA1/2 and KEA3 [46] act as osmosensors induced under osmotic stress – Ca_{2+} elevation. One of the plasma membrane receptors, known as plasma membrane receptor-like kinase FERONIA, interacts with the pectin component of the cell wall to maintain the integrity of the cell wall during salt stress [46, 48]. Calcium-permeable Cyclic nucleotide-gated ion channels are phosphorylated by FER and BAK1 and control calcium signaling. Calmodulin (CaM) aids in the regulation of cellular calcium levels, which inhibits these channels [49, 50]. In addition to this, the plasma membrane proteins leucine-rich-repeat receptor kinase and Hydrogen peroxide-induced Ca_{2+} increases 1 (HPCA1) detect the buildup of ROS, which is crucial for the activation of Ca_{2+} signaling during stomatal closure in response to stress stimuli [51].

5.4.2 MECHANISM ADAPTED BY PLANTS IN RESPONSE TO SALT-STRESS

Ionic balance, osmotic homeostasis, phytohormone signaling, cytoskeletal functions, and maintenance of cell wall structure are some of the mechanisms employed by plants in response to salt stress.

5.4.2.1 MAINTENANCE OF AN IONIC BALANCE

The plasma membrane of plant cells has localized Na_+/H_+ antiporters that play an important role in managing the higher concentrations of Na_+ ions in cells beyond a tolerance level [37]. Some regulatory pathways modulate the Na_+/H_+ antiporter activity to regulate the ionic balance during salt stress [43]. In high salinity circumstances, the cell's accumulation of Na_+ and Ca_{2+} results in the regulation of ion transport in plants. The salt excessively sensitive (SOS) pathway is activated by a rise in intracellular Ca_{2+} concentration and lessens salt stress by exporting Na_+. The SOS pathway is a complex of numerous regulatory molecules, including the calcium sensors SOS3 and SCaBP8, the protein kinase SOS2, the Na_+/H_+ antiporter SOS1, and the protein kinase SOS2 (SOS3-like calcium-binding protein 8) [53]. The calcium sensors SOS3/SCaBP8 perceive high intracellular calcium levels and transduce it to the downstream serine/threonine protein kinase and activate SOS2, which phosphorylates SOS1. Activation of SOS1 increases the Na_+/H_+ antiporter activity exchanging Na_+ ions and

H_+ ions through an antiport transport [38]. Thus, the SOS pathway is responsible for maintaining the Na_+ homeostasis and preventing the accumulation of Na_+ to toxic levels within a cell by transporting excess Na_+ from the cytosol.

To maintain a high K_+/Na_+ ratio, plants have to modulate Na_+/K_+ homeostasis activity [43]. Down-regulation of this activity leads to K_+ deficiency mainly due to salt stress or mutations. Voltage-gated channels, transporters, and their regulators are responsible for the regulation, absorption, translocation, and release of K_+ ions at cellular and whole-plant levels [55]. One of the major contributors to K_+ uptake is the *Arabidopsis* K transporter (AKT1), which is inhibited under salinity stress [55]. Certain proteins and kinases, such as calcineurin B-like (CBL) proteins and CBL interacting protein kinases (CIPKs), interact with AKT1 and activate it. AKT1 is phosphorylated and activated by CIPK23, thus increasing the uptake of K_+ under deficient conditions [56, 57]. Furthermore, to enhance the activity of AKT1, two CBL proteins, namely CBL1 and CBL9, activate the phosphorylation of CIPK23 on AKT1 [58]. Some negative modulators of AK1 decrease its activity through dephosphorylation or competitive binding, such as protein phosphatase 2C and CBL10, respectively [59]. The circadian clock, maintained by several proteins in plants, is a key element responsible for regulating salt stress, as demonstrated in *Oryza sativa* [60, 61].

Ionic balance is also regulated by aquaporins localized in the membrane responsible for root water uptake, stomatal closure, and leaf water transpiration. Aquaporin TdPIP2;1 found in wheat retains a low Na_+/K_+ ratio under salt stress conditions. Hence, improving the salt stress tolerance of wheat plants [62].

5.4.2.2 MAINTENANCE OF OSMOTIC HOMEOSTASIS

In plants, regulation of gene expression and activation of osmolyte biosynthesis enzymes are mediated through the osmotic signaling pathway in response to osmotic stress [63]. Plant cells under salinity stress, water deficiency, and ionic imbalance develop osmotic stress followed by shrinkage of the plasma membrane, alteration in the structure of the cell wall, and reduced turgor pressure [43]. Under salt stress, several osmolytes such as proline, polyols, and sugar accumulate and act as signaling molecules to accumulate ABA, regulate gene expression, promote plant growth, and participate in the reduction of the osmotic potential in the cytosol [64]. An increase in the levels of mitogen-activated protein kinases (MAPK) involved in ROS homeostasis leads to an increase in the accumulation of osmolytes [65, 66]. MPK3 and MPK6 induced by salt stress activate salt tolerance 1, receptor-like kinases, and also sense the increased concentration of cytosolic Ca_{2+}. Under salinity stress, the formation of reactive oxygen species (ROS) increases, and enzymes such as ascorbate peroxidases and catalase scavenge the ROS to develop tolerance towards salinity. It has been seen that increased expression of ascorbate peroxidases and glutathione S-transferase helps in salt tolerance in rice and *Arabidopsis* [67, 68].

5.4.2.3 PHYTOHORMONE SIGNALING – HELPS IN MEDIATING SALT TOLERANCE IN PLANTS

Plants produce two types of phytohormones: those that promote growth, such as auxin, gibberellins, cytokinins, brassinosteroids, and strigolactones, and those that respond to stress, such as ABA, ethylene, salicylic acid, and jasmonate (JA). Additionally, important to the plant's ability to respond to stress are brassinosteroids and strigolactones [69]. ABA, a 15-carbon isoprenoid, plays the role of a secondary mediator to activate kinase cascade reactions and gene expression for osmoregulation in a plant salt stress response [70]. The central component in the ABA signaling pathway is sucrose nonfermenting-1-related protein kinase 2s (SnRK2s), inhibited by phosphatases in the absence of ABA [71]. Under salt stress, ABA builds up and causes gene overexpression and ABA signaling pathway activation [72].

In reaction to salt stress, the other stress hormone, brassinosteroids (BR), builds up and uses osmoregulation to keep the ionic homeostasis in check. They reduce ROS levels and activate genes that respond to stress. By lowering cytosolic Na_{+} levels and enhancing K_{+} absorption through exogenous injection, they also demonstrate excellent salt tolerance [73]. They can positively regulate salt stress under the proper application of BR genes in transgenic plants. Genes such as the tomato BZR homolog gene, SlBZR1, have proven to positively regulate salt stress and up-regulate the expression of multiple stress-related genes. They also interact with other hormones such as ABA and cause the anti-stress effect.

5.4.2.4 MAINTENANCE OF CYTOSKELETON AND CELL WALL – THEIR FUNCTIONS AND REGULATIONS IN RESPONSE TO SALT-STRESS IN PLANTS

Actin filaments and microtubules form the cytoskeleton of plant cells. The cell vesicle trafficking, maintenance of cell shape, cell movement, growth, and response toward external stress stimuli are mediated by the cytoskeleton [63]. The main modulating dynamic events of forming a cytoskeleton structure are nucleation, polymerization, depolymerization, formation of bundles by crosslinking, and growth/shrinkage [75]. Cortical microtubules and actin filaments are depolymerized and destabilized in response to salt stress. These events lead to tolerance towards salt stress in plants [76, 77]. During calcium signaling in response to increased Ca_{2+} ion concentration, the cytoskeleton plays an important role in its upstream and downstream regulation [78]. Cortical granules in response to salt stress depolymerize and release Ca_{2+} ions into the cytosol. Actin nucleation factor is known as actin-related protein 2/3 (Arp 2/3) complex that regulates the integrity of mitochondria for the release of Ca_{2+} in salt stress. In *Arabidopsis* root, actin disorganization elicits a response towards salt stress via the production and accumulation of ROS [77].

Repression in cell expansion and division due to salt stress in plants destabilizes the cell shape and its functions, which positively affects the salt stress response in plants. The cell wall helps to maintain the shape of the cell even under variant turgor pressure, caused by water deficiency under saline conditions [80]. The cell wall, being the first line of defense, with the help of salt sensors found on the plasma membrane, acts as an early sensor for salt stress. FER, a salt sensor, acts as a receptor kinase and binds with RALF peptides, which along with leucine-rich repeat extensins (LRX), regulate downstream signals for promoting plant growth even under cell wall integrity damage [81]. The plant cell wall consists of polysaccharides such as lignin, pectin, and cellulose regulated by transcription factors NAC domain, Homeobox HD-ZIP Class III (HD-ZIPIII), and phytohormones (ABA) under salt stress [82, 83]. SnRK2 kinases interact with transcription factors NAC-domain promoting secondary cell wall thickening and lignification in response to salt stress tolerance [84].

5.5 IMPACT OF SALINITY ON PLANT GROWTH AND SOIL QUALITY

The major abiotic stress on the plant is salinization, which contributes to the loss of vegetative crop production and yields in the agricultural sector, mainly due to irrigation. It is estimated that around 1 million hectares of land in the European Union are a major cause of desertification [85]. The rate of salinization is expected to accelerate due to several factors such as climate change, the use of low-quality water for irrigation, poor drainage, the use of groundwater, etc. Soluble salts in the soil are easily absorbed by plants, and excessive accumulation of these salts suppresses plant growth, affecting the soil's physicochemical properties and leading to soil erosion, low productivity, and low economic returns [86]. Several stages of plant growth are affected by soil salinity, which includes seed germination, food storage, photosynthesis, and vegetative and reproductive development, thereby inhibiting the development of the microsporangia and megasporangia of a plant [87]. A decrease in chlorophyll content, leaf surface area, and stomatal conductance is also affected by soil salinity. Ionic imbalance and increased osmotic pressure beyond 0.2 MPa due to salinization impact plant nutrient uptake and turgidity of the cell, respectively. Hence, leading to dehydration, impairment in growth, and ultimately cell death [88]. It has been reported that soil salinity has profound adverse effects on plant growth during their reproductive stage, e.g., in the wheat plant, reduction and delayed emergence in the spikelets per spike have been reported due to salinization resulting in poor grain yield at 100–175 mM NaCl conc [89]. Cyclin and cyclin-dependent kinase activity and expressions are greatly reduced under salt stress conditions, thus arresting the normal cell cycle that affects the meristematic growth, enzymatic activity, mitosis, DNA, and RNA synthesis required for the development of a plant [90].

Salinization also affects the soil microbial community and enzymatic activities in two ways: increasing the osmotic potential of soil water and ionic imbalance. Due to the increase in the osmotic potential of soil water, plasmolysis of microbial cells and root cells takes place, which draws water out of the cells, ultimately leading to cell death. To combat this situation, plants and microbes accumulate osmolytes, mainly proline, glycine, and K_+ ions, to prevent plasmolysis. This whole process of accumulating osmolytes is quite energy-consuming and therefore hampers both the growth and activity of plants and microorganisms [91]. Microbial activity, microbial biomass, and the structure of a microbial community are the key features that govern soil quality. These features improve the fertility and nutrient content of the soil, which however gets adversely affected by prolonged and unmanageable soil salt stress [92]. Increased Na_{+} Cl_{-} ion concentration interferes with the physicochemical properties of soil, whereas in plants it leads to tissue necrosis and chlorosis. Salinization renders land unproductive, infertile, and deserted, with soil texture variation, differed soil appearance, and varied soil coloration. According to several studies, soil texture regulates the makeup of the soil microbial population, which in turn affects the detrimental effect of salt on C mineralization. A comparison between fungi and bacterial response towards salt stress has reported that soil bacterial composition is highly affected under salt stress and low water potential [93]. Proper management of crop yield and the use of current advancements in techniques for irrigation need to be employed to prevent the desertification of agricultural land globally. The use of transgenic parents, controlled irrigation, good quality water, and incorporation of NPs and nano-BCs in soil have proven to be an effective approach to mitigate the effects of soil salinization globally.

5.6 NANO AND NANO-BIOCHAR AS A SOIL AMELIORANT – REDUCING SALINITY STRESS

5.6.1 ROLE OF NANOPARTICLES IN ALLEVIATING SALT STRESS IN PLANTS

Nanotechnology has gained fame in several sectors such as medicine, electronics, agriculture, industries, etc. It is believed that with the advent of nanotechnology, the introduction of nano-sized particles in the agricultural and environmental remediation sector has proven to play a key role in enhancing soil productivity, nutrient content, and crop yield, as well as alleviating salt stress in plants. Nano-particles (NPs) possess unique physicochemical properties to improve plant metabolism. These nano-sized engineered particles can enter a plant cell and act as a vector for the transport of DNA and other chemical molecules inside a plant cell for specific gene manipulation, regulation, and expression. NPs have proven to increase the light-harvesting capacity of plants through the incorporation of carbon nanotubes, an artificial antenna in chloroplasts to catch wavelengths of light usually not in the typical range. Apart from this, carbon nanotubes help in seed germination, proper development, and growth of a plant [94, 95]. Prominent morphological and physiological alterations are observed in plants after interaction with NPs, whose chemical composition, particle size, dosage, surface area, and reactivity are major properties for an efficient outcome [96].

5.6.1.1 SILICA NANOPARTICLE

The second-most abundant material in the Earth's crust, silica, is essential for improving plant tolerance to biotic and abiotic stress (salinity stress, metal toxicity, drought). Plant metabolism, growth, and development are increased by silica nanoparticles derived from natural sources, which also enhance plant photosynthesis [97]. Silica NPs increase the water absorption capacity of plants by decreasing evapotranspiration through stomata. Silicon regulatory genes LSi1 and LSi2 are up-regulated in jasmonic acid (JA)-phytohormone signaling under saline conditions, initiating osmolyte production and antioxidant defense system. Alsaeedi et al. [98] studied that the seeds and seedlings of common beans grown under increased NaCl concentration showed improved growth, sped-up seed germination, and improved shoot and root length after treatment with SiNPs, thus relieving the detrimental effects of NaCl on plant growth [98].

5.6.1.2 CERIUM OXIDE NANOPARTICLES

CeO_2NPs possess unique catalytic and optical properties which make them efficient for the production of microelectronics, catalysts, sunscreen, and several polishing agents. CeO_2NPs are also added to petrol and diesel to enhance fuel ignition proficiency and reduce diesel soot emissions, respectively. Moreover, these particles play a major role in enhancing plant health under certain external exposure conditions both negatively and positively [99]. The incorporation of 10 g L_{-1} CeO_2NPs during the irrigation of plants has been reported to improve tomato yield and productivity, thus reducing both oxidative and salinity stress in the environment. Canola plants treated with CeO_2NPs showed improved plant biomass, photosynthetic capacity, and salt tolerance by modifying root anatomy and physiology. Plants under high salt concentration develop a short root apoplastic barrier that allows the transport of Na_+ towards the shoot and prevents the accumulation of Na_+ ions in roots in the presence of CeO_2NPs for the better physiological performance of a plant [100].

5.6.1.3 ZINC OXIDE NANOPARTICLES

Due to its antibacterial qualities, zinc oxide (ZnO) is frequently used in food packaging and pharmaceuticals. In addition, Zn, a micronutrient, is utilized in the manufacturing of insecticides and fertilizers. Additionally, Zn plays a crucial role as a co-factor in several biocatalytic enzymes, including hydrolases, oxidoreductases, transferases, and ligases [101]. Based on these peculiar qualities, ZNPs have been employed in various agricultural sectors for the development and growth of plants. Several experiments on peanut, wheat, onion, and soybean reported that an increase in the concentration of ZNPs resulted in delayed or impaired seed germination whereas a decrease or lower concentration of ZNPs had beneficial effects on seed germination. Additionally, it dramatically improves microbial population, protein, and chlorophyll synthesis, and root and shoots development. The addition of ZNPs to MS media encourages proline synthesis, somatic embryogenesis, and plantlet regeneration. ZNPs also encourage peroxidase, catalase, and superoxide dismutase activities to resist biotic stress [102].

5.6.1.4 NITRIC OXIDE-CHITOSAN NANOPARTICLES

Different effects of abiotic stresses on plants are mitigated by the action of a gaseous signaling molecule known as nitric oxide. NO is responsible for modulating gene expression, enhancing several prominent protein functions, chlorophyll biosynthesis, osmolyte accumulation in cells, efficient H_+-ATPase activity, and strengthening the antioxidant defense system in plants. In the agricultural and environmental sectors, the negative impact of salinity stress on plant development and growth is reduced by the exogenous application of NO [103]. Due to its volatile nature, direct administration of NO in plants is a challenging operation. However, chitosan nanoparticles can be successfully used for a variety of biological applications, such as antibacterial activity against several resistant microorganisms. S-nitrosomercaptosuccinic acid, a low-molecular-weight NO donor used to promote salt tolerance in plants experiencing salt stress, is a component of CS-NPs. Because of this integration, NO can be released continuously, reducing the negative effects of salinity on plant growth and photosynthesis [104].

5.6.1.5 SILVER (AG) AND IRON (FE) NANOPARTICLES

The use of AgNPs and FeNPs in the agricultural sector has provided a sustainable approach toward the improvement of crop yield. The AgNPs and FeNPs positively help in alleviating salinity stress and possess great potential in ameliorating salinity stress in plants and enhancing the salt tolerance capacity of plants. However, the molecular approach of these NPs towards salinity tolerance in plants further needs to be evaluated and studied in detail.

By boosting the phenolic and flavonoid chemicals in the root region of plants, which affect how well plants respond to saline stress, FeNPs upregulate the essential antioxidative enzymes. Numerous essential cellular processes, including photosynthesis, respiration rate, and the manufacture of chlorophyll, are aided by it. By boosting the expression of genes in both the smaller and bigger enzyme subunits involved in photosynthesis, it improves photosynthesis in plants. Whereas AgNPs are essential for boosting the chlorophyll content and fluorescence characteristics of plants exposed to salinity [105, 106].

Based on the above-mentioned parameters of various nanoparticles, one of the major benefits of NPs includes the increased photosynthetic capacity to improve crop yield and plant tolerance under harsh saline conditions.

5.6.1.6 ROLE OF MICRONUTRIENTS CU-MN NPS IN ALLEVIATING SALT-STRESS

Copper (Cu) as a micronutrient plays a significant role in minimizing the detrimental effects of salinity on plants. It helps to improve water relations, photosynthesis, and nutrition and up-regulates the antioxidant defense. Moreover, it helps to increase levels of osmoprotectants and amino acids. The impact of Cu-NP on tomato was examined under salt stress in a study by Hernández-Hernández et al. [107] who discovered that by increasing the expression of the SOD and jasmonic acid (JA) genes, reduction of ionic and oxidative stressors was found along with increased growth of tomato [107]. Therefore, Cu-NPs could successfully boost salinity tolerance by triggering the antioxidant defense system and the jasmonate octadecanoid pathway.

Under abiotic stress, it was also discovered that Manganese (Mn) can also support maintaining an ideal rate of photosynthesis. *Vigna radiata* plants supplemented with FeNPMn under salt stress showed improved membrane stability index, chlorophyll content, and nitrate reductase activity [108]. On the other hand, there is no information on how to increase plants' tolerance to salt stress caused by photosynthesis. Future research should focus on this area to better understand the potential and novel functions of Mn and Cu-NPs in affecting salt stress tolerance.

5.7 PHYTOTOXICITY CONCERNING THE USE OF NANOPARTICLES

The implementation of nanotechnologies in the agricultural sector has raised concerns regarding the bioavailability and toxicity of nanoparticles. Improper use of NPs has negative effects on plant growth. Furthermore, the toxicity of NPs reduces the formation of proteins, pigments, and several enzymes required for the proper development of a plant. Various morphological, physiological, genetic, and proteomic responses of plants are affected due to the high concentrations of NPs mentioned in Table 5.3. NPs' phytotoxicity results in oxidative stress via ROS production. Therefore, plant cells create a variety of enzymatic and non-enzymatic antioxidants to reduce oxidative stress, including CAT, POD, SOD, thiols, phenolic compounds, and ascorbate. Methalothinan and phytochelatins are among the several other compounds produced in response to NP stress to safeguard a plant against the adverse effects of oxidative stress [109, 110].

5.7.1 ROLE OF NANO-BIOCHAR IN ALLEVIATING SALT STRESS

Increased soil contamination has become a great threat to the agriculture sector. Especially the toxicity of arsenic metal (As), along with the high concentration of salt in the soil, has shaken food security and public health globally. Nano-BCs, being highly porous, having selective functional groups, and compact nature, increase the adsorption rate of minerals in plants grown under saline conditions. In halophytes, e.g., Quinoa plants, phytotoxicity due to the combined effect of As and salinity doesn't provide desirable crop results. Hence, the incorporation of BC-doped silicon nanoparticles proved to mitigate the saline effects and resulted in high grain yield and chlorophyll content individually as well as synergistically [111]. Artificially engineered BCs or rice BCs combined with Iron oxide NPs alleviate the soil As and salinity stress in plants. A significant portion of As was adsorbed on the surface of the nano-BCs conjugated with Iron oxide. This combination resulted in the bioavailability of nutrients and minerals and improved the internal vascular structure of plants. This conjugated mechanism has proven to be one of the best combinations to alleviate and mitigate the salinity as well as arsenic stress in agricultural sectors.

Further, the application of nano-BC reduces the oxidative and antioxidant enzymatic stress on plants. Hence, it also balances the pH of the soil, enhances calcium and potassium ion uptake, and even maximizes the effectiveness of photosystem II and the production of numerous photosynthetic pigments [112]. In soybean plants, drought and salinity significantly decrease the flowering time, water use efficiency, photosynthesis, stomatal conductance, root growth, and absorption rate. It also results in lowered crop yield by 24–50% and biomass production. Thus, it was reported that the addition of nano-BC enhanced crop yield production by 3.1–14.8% and increased the water use efficiency under drought conditions [113]. Salinity, being a crucial environmental factor, restricts plant growth and crop yield, affecting agricultural sustainability. Therefore, the effective use of nano-BC in agriculture has improved crop quality and yield tremendously in the past few years. Significantly, biochar-based Nanocomposites (BNC) in agriculture have given a ray of positive hope by improving the grain/crop yield. In safflower plants, positive and promising effects were seen when a combination of biochar and nanocomposites including magnesium (Mg) and manganese (Mn) were used. Under 6–12 dS/m saline conditions, absorption rate, plant growth and development, nutrient uptake, and sodium exchange were hampered. With the administration of BNC, a decrease in the negative effects of salinity was observed, and significantly improved crop yield was achieved [114].

The world cultivates eggplant because it is a rich source of vitamins and minerals, has a low production cost, and has a high antioxidant value. These plants are moderately sensitive to salt stress under the threshold range of 1.1–1.5 dS/m. Several modifications and adaptations have been mastered by the plant to develop efficient crop cultivation practices and breeding strategies. Under salinity stress, changes in the root and shoot parameters of the plant were observed. As a result, factors that affect photosynthesis, transpiration, respiration, and the movement of minerals and nutrients within plants also have an impact on the quality and production of the crop. The use of hardwood and softwood biochar produced from oak and pine wood increases the efficacy and efficiency of the plant tremendously. The addition of biochar showed a positive impact on the soil quality and multifold expansion in the water-holding capacity of the soil [115]. *Solanum tuberosum* is the widely cultivated vegetable crop and most consumed crop after wheat, rice, and maize globally, commonly known as potato. Nano-BC produced from rice husk, pine woods, etc., combined with inorganic fertilizers increases the crop yield of potatoes under saline conditions and also influences the nitrogen content of soil positively [116].

Nano-biochar reduces the salt stress in plants due to its large surface area, which significantly increases the adsorption rate of Na_+ ions on its surface. Wheat plants grown under increased salt concentration reported better yield and growth in the presence of nano-BC. The physicochemical and biological properties of nano-BC have been proven beneficial for enhancing soil quality and crop yield. The interaction of nano-BC with several soil components eliminates the toxic metal ions, Na_+ ions, inorganic compounds, etc., that are responsible for the impairment of plant growth and development. Therefore, it interferes with the smooth physiological functioning and development of plants and helps to determine the different effects of nano-BC produced from various raw materials. Hence, the selection of better raw materials and production processes determines the quality, efficiency, characteristics, and biological and physicochemical properties of a particular nano-BC in mitigating the salinity effects on plants in the environment under salt stress [117].

5.8 METHODS INVOLVED IN THE AGRICULTURAL SECTOR TO EVALUATE THE MITIGATING EFFECTS OF SALINITY STRESS IN PLANTS THROUGH NANO AND NANO-BIOCHAR TREATMENT

Several methods have been employed to incorporate and evaluate the effect of NPs in plants under saline conditions which are briefly described in Tables 5.4 and 5.5.

5.9 FUTURE SCOPE

Numerous functions and techniques can be used to lessen the restrictions and potential risks associated with nano-BC particles. Thus, Nano-BC provides

TABLE 5.3 Phytotoxic Effect of Some Prominently Used Nanoparticles (NPs) in an Agricultural Sector

SL. No.	Stages	NPs	Experimental Plant	Toxic Effect	References
1.	Morphological	• ZnO NPs • CuO NPs	• *Arabidopsis thaliana* • *Fagopyrum esculentum*	• Decrease in plant biomass. • Root and shoot elongation get inhibited. • In the root region, abnormal cell growth occurs.	[102]
2.	Genetic	• ZnO NPs • AgNPS • CuNPs	• *Allium cepa* • *Allium cepa* • *Raphanus sativus*	• This leads to chromosomal aberrations and decreases mitosis. • Chromosomal breaks, impaired cell division, and chromosomal bridge formation. • This leads to DNA damage.	[103, 106, 108]
3.	Proteomic	• AgNPs	• *Triticum sativum* • *Oryza sativa*	• Decreased protein synthesis: Cell wall, mitochondrial ETC. • Accumulation of protein precursors which inhibit normal cell reproduction.	[106]

TABLE 5.4 NPs and Methods Involved to Alleviate Salinity Stress in Plants

SL. No.	Methods	Experimental Plants	NPs-Conc mg/L	Salt Conc (mM)	NPs Positive Impact on Plants	References
1.	Foliar spraying	**Lycopersicon esculentum* **Dracocephalum moldavica*	Fe-NPs 50–100 mg/L	50–150 mM	Increase in enzymatic and non-enzymatic activities along with improvement in morphological and photosynthetic activity.	[118]
2.	Priming	**Teriticum aestivum*	ZnO-NPs 50 mg/L	200 mM	Enhances trapped energy flux and ETC flux activation of antioxidant system and biosynthesis of sucrose.	[119]
3.	Nutrient medium	**Ocimum basilicum*	Carbon-NPs 50–100 mg/L.	50–100 mM	Increase in chlorophyll content.	[120]
4.	Growth medium	**Lycopersicon esculentum*	Si-NPs 250–500 mg/L.	50 mM	Maintenance of chlorophyll, vitamin C, and GSH content.	[121]
5.	Soil	**Solanum tuberosum L*	Zn, B, Si, and zeolite NPs Range- 2–15 mg/L.	4.2 dS/m	Improvement in morphological traits of a plant, improved adsorption, and concentration of nutrients.	[122]

TABLE 5.5 Nano-BC and BNCs Involved in Alleviate Salinity Stress in Plants

SL. No	Nano-Biochar Treatments	Raw Material	Plant	NaCl Conc mM	Positive Impact	References
1.	Hardwood biochar-softwood biochar.	Oak and pine	Eggplant	50–100 mM	Significantly improved soil properties, and enhanced water-holding capacity. Increases photosynthetic rate.	[115]
2.	Nano-BC	Corn stalks	Soyabean	50–100 mM	Balances the pH of soil. Increases calcium and potassium ion uptake. Maximizes the efficiency of photosystem-II. Increase synthesis of several photosynthetic pigments.	[113]
3.	Nano-BC	Rice, wheat husks	Potato	25–50 mM	Increased shoot biomass, root length, crop yield, Increased total nitrogen content.	[116]
4.	Nano-BC + nano-composites (MgO, MnO)	Pinewood, stalks, rice, and wheat husks	Safflower	50–150 mM	Increases chlorophyll content (a & b), efficient nutrient uptake and water absorption, Improved growth of root and shoot.	[114]
5.	Nano-BC + Si	Oaktree, husks, grass.	Quinoa plant	40–200 mM	High grain yield and chlorophyll content individually as well as synergistically.	[111]

a wide range of alternatives in agricultural activities in comparison to its bulk equivalent [79, 123–134]. Numerous studies have so far demonstrated appreciable modifications in soil physicochemical and biological parameters as a result of the addition of nano-BC. Because of their small size and increased surface area, nano-BCs have the potential to remove organic and inorganic pollutants from the environment, improving soil quality and crop performance. To encourage plant growth, microbial variety, and soil biological processes, it is possible to transfer the nutrients associated with nano-BC to native soil plants and microbes. Therefore, further fieldwork would be required before nano-BC is suggested for application in agriculture. Additionally, studies are being conducted to determine whether the buildup of nano-BC in produced crops has any negative consequences on the environment. In addition to acting as a habitat for bacteria, the bigger SSA of nano-BC is an essential site for the absorption of nutrients like phosphorus.

Nano-particles possess unique properties that have led to several novel applications in a range of scientific fields. NP effects on plants vary from plant to plant and depend on their size, concentration, and delivery mode. The understanding of the physiological, biochemical, and molecular mechanisms of plants concerning NPs is still in its infancy; additionally, in-depth investigations are needed. Future research is also predicted to examine the operation of NPs, their interactions with biomolecules, and their effects on how plants control the expression of their genes. The use of NPs, one of the most promising strategies to enhance growth and plant performance under salt stress, has received a lot of attention in recent years [42, 47, 52, 54, 74].

5.10 CONCLUSION

Native nano-BC has several drawbacks that prevent it from being used directly in the environment, including low stability and difficulty isolating nanoparticles, easy mobilization, uptake, accumulation, and nature of toxicity concerning very small particles (nano to micron). The size distribution of nano-BC may be influenced by several variables, including the size of the source material, physical grinding or milling, and the aging procedure.

Nanomaterials have a significantly bigger surface area when compared to the same mass of material created in a larger form. Nanoparticles can increase a material's chemical reactivity while simultaneously changing its strength or electrical properties. The surface-to-volume ratio of nanoparticles is high, which increases their reactivity and potential biological activity. Nanoparticles regulate salt tolerance in a variety of plants by affecting hormone concentrations, antioxidant enzyme activity, ion homeostasis, gene expression, and defense mechanisms. Nanotechnology, which is the study and use of these unusually small objects, focuses on small-scale materials between 1 and 100 nm in at least one of their dimensions (length, height, and breadth). These nanomaterial properties, such as electrical conductivity, magnetism, chemical reactivity, optical effects, and physical strength, differ from those of bulk materials due to their smaller size and greater surface area. There are numerous ways to apply NPs, including irrigation, hydroponic substrate, direct injection, foliar spraying, and priming. When NPs are exposed to plants, they enter the root junction and wound regions of the plant, pass through the cell membrane and cell wall of the root epidermis (e.g., through endocytosis, carrier proteins, plasmodesmata, or pore formation), and then move through one of the two pathways of symplast or apoplast to enter the plant's vascular bundle (xylem). The application methods, size, concentration, and environment are the main factors that impact the effective adsorption of NPs after foliar application. This efficient technique is utilized in different sectors of agricultural nanotechnologies, such as nano-herbicides, nano-fungicides, nano-fertilizers, nano-pesticides, nanosensors, etc., to raise the number and quality of various goods as well as solve some issues.

KEYWORDS

- **agricultural sector**
- **calcineurin B-like**
- **calmodulin**
- **electrical conductivity**
- **ionic imbalance**
- **leucine-rich repeat extensins**
- **mitogen-activated protein kinases**
- **nano-biochar**
- **nanoparticle**
- **osmotic potential**
- **reactive oxygen species**
- **salinity stress**

REFERENCES

1. Goswami, L., Kushwaha, A., Singh, A., Saha, P., Choi, Y., Maharana, M., Patil, S. V., & Kim, B. S. (2022). Nano-biochar as a sustainable catalyst for anaerobic digestion: A synergetic closed-loop approach. *Catalysts, 12*, 186. https://doi.org/10.3390/catal12020186.
2. Rajput, V. D., Minkina, T., Ahmed, B., Singh, V. K., Mandzhieva, S., Sushkova, S., Bauer, T., Verma, K. K., Shan, S., van Hullebusch, E. D., & Wang, B. (2022). Nano-biochar: A novel solution for sustainable agriculture and environmental remediation. *Environmental Research, 210*, 112891. https://doi.org/10.1016/j.envres.2022.112891.
3. Lomas, J., & Levin, J. (1979). Irrigation. In *Agrometeorology* (pp. 217–264). Springer.
4. Song, T., Xu, H., Sun, N., Jiang, L., Tian, P., Yong, Y., Yang, W., Cai, H., & Cui, G. (2017). Metabolomic analysis of alfalfa (*Medicago sativa* L.) root-symbiotic rhizobia responses under alkali stress. *Frontiers in Plant Science, 8*, 1208. https://doi.org/10.3389/fpls.2017.01208.
5. Zhang, L., Xiang, Y., Jing, Y., & Zhang, R. (2019). Biochar amendment effects on the activities of soil carbon, nitrogen, and phosphorus hydrolytic enzymes: A meta-analysis. *Environmental Science and Pollution Research, 26*, 22990–23001. https://doi.org/10.1007/s11356-019-05145-8.
6. Zhao, S., Zhang, Q., Liu, M., Zhou, H., Ma, C., & Wang, P. (2021). Regulation of plant responses to salt stress. *International Journal of Molecular Sciences, 22*(9), 4609. https://doi.org/10.3390/ijms22094609.
7. Quan, R., Lin, H., Mendoza, I., Zhang, Y., Cao, W., Yang, Y., Shang, M., Chen, S., Pardo, J. M., & Guo, Y. (2007). SCABP8/CBL10, a putative calcium sensor, interacts with the protein kinase SOS_2 to protect Arabidopsis shoots from salt stress. *The Plant Cell, 19*, 1415–1431. https://doi.org/10.1105/tpc.106.042291.
8. Gong, Z. (2021). Plant abiotic stress: New insights into the factors that activate and modulate plant responses. *Journal of Integrative Plant Biology, 63*, 429–430. https://doi.org/10.1111/jipb.13079.
9. Zhu, J. K. (2002). Salt and drought stress signal transduction in plants. *Annual Review of Plant Biology, 53*, 247–273. https://doi.org/10.1146/annurev.arplant.53.091401.143329.
10. Haider, F. U., Wang, X., Farooq, M., Hussain, S., Cheema, S. A., Ain, N. U., Virk, A. L., Ejaz, M., Janyshova, U., & Liqun, C. (2022). Biochar application for the remediation of trace metals in contaminated soils: Implications for stress tolerance and crop production. *Ecotoxicology and Environmental Safety, 230*, 113165. https://doi.org/10.1016/j.ecoenv.2022.113165.
11. Ameloot, N., Graber, E. R., Verheijen, F. G. A., & De Neve, S. (2013). Interactions between biochar stability and soil organisms: Review and research needs. *European Journal of Soil Science, 64*, 379–390. https://doi.org/10.1111/ejss.12064.
12. Liu, W., Li, Y., Feng, Y., Qiao, J., Zhao, H., Xie, J., Fang, Y., Shen, S., & Liang, S. (2020). The effectiveness of nano biochar for reducing phytotoxicity and improving soil remediation in cadmium-contaminated soil. *Scientific Reports, 10*(1), 858. https://doi.org/10.1038/s41598-019-57128-4.

13. Oni, B. A., Oziegbe, O., & Olawole, O. O. (2019). Significance of biochar application to the environment and economy. *Annals of Agricultural Sciences, 64*, 222–236. https://doi.org/10.1016/j.aoas.2019.09.001.
14. Guo, F., Bao, L., Wang, H., Larson, S. L., Ballard, J. H., Knotek-Smith, H. M., Zhang, Q., Su, Y., Wang, X., & Han, F. (2020). A simple method for the synthesis of biochar nanodots using a hydrothermal reactor. *MethodsX, 7*, 101022. https://doi.org/10.1016/j.mex.2020.101022.
15. Jiang, C., Bo, J., Xiao, X., Zhang, S., Wang, Z., Yan, G., Wu, Y., Wong, C., & He, H. (2020). Converting waste lignin into nano-biochar as a renewable substitute for carbon black for reinforcing styrene-butadiene rubber. *Waste Management, 102*, 732–742. https://doi.org/10.1016/j.wasman.2019.11.034.
16. Ramanayaka, S., Tsang, D. C. W., Hou, D., Ok, Y. S., & Vithanage, M. (2020). Green synthesis of graphitic nano biochar for the removal of emerging contaminants in aqueous media. *Science of the Total Environment, 706*, 135725. https://doi.org/10.1016/j.scitotenv.2019.135725.
17. Nath, B. K., Chaliha, C., & Kalita, E. (2019). Iron oxide permeated mesoporous rice-husk nano biochar (IPMN) mediated removal of dissolved arsenic (As): Chemometric modeling and adsorption dynamics. *Journal of Environmental Management, 246*, 397–409. https://doi.org/10.1016/j.jenvman.2019.05.114.
18. Li, R., Zhang, Y., Deng, H., Zhang, Z., Wang, J. J., Shaheen, S. M., Xiao, R., et al. (2020). Removing tetracycline and Hg (II) with ball-milled magnetic nano biochar and its potential on polluted irrigation water reclamation. *Journal of Hazardous Materials, 384*, 121095. https://doi.org/10.1016/j.jhazmat.2019.121095.
19. Naghdi, M., Taheran, M., Pulicharla, R., Rouissi, T., Brar, S. K., Verma, M., & Surampalli, R. Y. (2019). Pine-wood derived nano biochar for removal of carbamazepine from aqueous media: Adsorption behavior and influential parameters. *Arabian Journal of Chemistry, 12*(8), 5292–5301. https://doi.org/10.1016/j.arabjc.2016.09.011.
20. Mahmoud, M. E., Abdelfattah, A. M., Tharwat, R. M., & Nabil, G. M. (2020). Adsorption of negatively charged food tartrazine and sunset yellow dyes onto positively charged triethylenetetramine biochar: Optimization, kinetics, and thermodynamic study. *Journal of Molecular Liquids, 318*, 114297. https://doi.org/10.1016/j.molliq.2020.114297.
21. Ardebili, S. M. S., Taghipoor, A., Solmaz, H., & Mostafaei, M. (2020). The effect of nano biochar on the performance and emissions of a diesel engine fueled with fuel oil-diesel fuel. *Fuel, 268*, 117356. https://doi.org/10.1016/j.fuel.2020.117356.
22. Mahmoud, E., El Baroudy, A., Ali, N., & Sleem, M. (2020). Spectroscopic studies on the phosphorus adsorption in salt-affected soils with or without nano-biochar additions. *Environmental Research, 184*, 109277. https://doi.org/10.1016/j.envres.2020.109277.
23. Mukhopadhyay, R., Sarkar, B., Jat, H. S., Sharma, P. C., & Bolan, N. S. (2021). Soil salinity under climate change: Challenges for sustainable agriculture and food security. *Journal of Environmental Management, 280*, 111736. https://doi.org/10.1016/j.jenvman.2020.111736.

24. Okur, B., & Örçen, N. (2020). Soil salinization and climate change. In *Climate change and soil interactions* (pp. 331–350). Elsevier. https://doi.org/10.1016/B978-0-12-818032-7.00012-6.
25. Sverdrup, H., & Warfvinge, P. (1988). Weathering of primary silicate minerals in the natural soil environment in relation to a chemical weathering model. *Water, Air, & Soil Pollution, 38*, 387–408. https://doi.org/10.1007/BF00248194.
26. Whipkey, C. E., Capo, R. C., Chadwick, O. A., & Stewart, B. W. (2000). The importance of sea spray to the cation budget of a coastal Hawaiian soil: A strontium isotope approach. *Chemical Geology, 168*, 37–48. https://doi.org/10.1016/S0009-2541(00)00187-X.
27. Rodríguez, A., Íguez, M., González Soto, L., Hernández Hernández, C., Jiménez Mendoza, C., Ortega González, M. J., Padrón, P. A., Torres Cabrera, J. M., & Vargas Chávez, G. E. (1993). Assessment of soil degradation in the Canary Islands (Spain). *Land Degradation & Development, 4*(1), 11–20. https://doi.org/10.1002/ldr.3400040104.
28. Di Liberto, V., Nuccio, P. M., & Paonita, A. (2002). Genesis of chlorine and sulfur in fumarolic emissions at Vulcano Island (Italy): Assessment of pH and redox conditions in the hydrothermal system. *Journal of Volcanology and Geothermal Research, 116*, 137–150. https://doi.org/10.1016/S0377-0273(02)00215-9.
29. Shrivastava, P., & Kumar, R. (2015). Soil salinity: A serious environmental issue and plant growth promoting bacteria as one of the tools for its alleviation. *Saudi Journal of Biological Sciences, 22*, 123–131. https://doi.org/10.1016/j.sjbs.2014.12.003.
30. Ivushkin, K., Bartholomeus, H., Bregt, A. K., Pulatov, A., Kempen, B., & de Sousa, L. (2019). Global mapping of soil salinity change. *Remote Sensing of Environment, 231*, 111260. https://doi.org/10.1016/j.rse.2019.111260.
31. Jamil, A., Riaz, S., Ashraf, M., & Foolad, M. R. (2011). Gene expression profiling of plants under salt stress. *Critical Reviews in Plant Sciences, 30*, 435–458. https://doi.org/10.1080/07352689.2011.605739.
32. Qadir, M., Oster, J. D., Schubert, S., Noble, A. D., & Sahrawat, K. L. (2007). Phytoremediation of sodic and saline-sodic soils. *Advances in Agronomy, 96*, 197–247. https://doi.org/10.1016/S0065-2113(07)96006-X.
33. Wichelns, D., & Qadir, M. (2014). Achieving sustainable irrigation requires executive management of salts, soil salinity, and shallow groundwater. *Agricultural Water Management, 157*, 31–38. https://doi.org/10.1016/j.agwat.2014.08.016.
34. Alexandre, C., Borralho, T., & Durao, A. (2018). Evaluation of salinization and sodification in irrigated areas with limited soil data: Case study in southern Portugal. *Spanish Journal of Soil Science, 8*, 102–120. https://doi.org/10.3232/SJSS.2018.V8.N1.01.
35. Su, T., Li, X., Yang, M., Shao, Q., Zhao, Y., Ma, C., & Wang, P. (2020). Autophagy: An intracellular degradation pathway regulating plant survival and stress response. *Frontiers in Plant Science, 11*, 164. https://doi.org/10.3389/fpls.2020.00164.
36. Horie, T., Karahara, I., & Katsuhara, M. (2012). Salinity tolerance mechanisms in glycophytes: An overview with the central focus on rice plants. *Rice, 5*, 11–18. https://doi.org/10.1186/1939-8433-5-11.

37. Zhu, J. K. (2002). Salt and drought stress signal transduction in plants. *Annual Review of Plant Biology, 53*, 247–273. https://doi.org/10.1146/annurev.arplant.53.091401.143329.
38. Van Zelm, E., Zhang, Y., & Testerink, C. (2020). Salt tolerance mechanisms of plants. *Annual Review of Plant Biology, 71*, 403–433. https://doi.org/10.1146/annurev-arplant-050718-100005.
39. Chen, Y., & Hoehenwarter, W. (2015). Changes in the phosphoproteome and metabolome link early signaling events to the rearrangement of photosynthesis and central metabolism in salinity and oxidative stress response in Arabidopsis. *Plant Physiology, 169*, 3021–3033. https://doi.org/10.1104/pp.15.01486.
40. Shumilina, J., Kusnetsova, A., Tsarev, A., Van Rensburg, H. C. J., Medvedev, S., Demidchik, V., Ende, W. V. D., & Frolov, A. (2019). Glycation of plant proteins: Regulatory roles and interplay with sugar signaling? *International Journal of Molecular Sciences, 20*, 2366. https://doi.org/10.3390/ijms20092366.
41. Gong, Z. (2021). Plant abiotic stress: New insights into the factors that activate and modulate plant responses. *Journal of Integrative Plant Biology, 63*(3), 429–430. https://doi.org/10.1111/jipb.13079.
42. Singh, A., Sengar, R. S., Shahi, U. P., Rajput, V. D., Minkina, T., & Ghazaryan, K. A. (2022). Prominent effects of zinc oxide nanoparticles on roots of rice (*Oryza sativa* L.) grown under salinity stress. *Stresses, 3*, 33–46. https://doi.org/10.3390/stresses3010004.
43. Park, H. J., Kim, W.-Y., & Yun, A. D.-J. (2016). A new insight into salt stress signaling in plants. *Molecular Cells, 39*(6), 447–459. https://doi.org/10.14348/molcells.2016.0083.
44. Pollard, R. T., Salter, I., Sanders, R. J., Lucas, M. I., Moore, C. M., Mills, R. A., Statham, P. J., et al. (2009). Southern Ocean deep-water carbon export enhanced by natural iron fertilization. *Nature, 457*(7229), 577–580. https://doi.org/10.1038/nature07716.
45. Zhang, S., Wu, Q.-R., Liu, L.-L., Zhang, H.-M., Gao, J.-W., & Pei, Z.-M. (2020). Osmotic stress alters circadian cytosolic Ca_{2+} oscillations and OSCA1 is required in circadian-gated stress adaptation. *Plant Signaling & Behavior, 15*(11), 1836883. https://doi.org/10.1080/15592324.2020.1836883.
46. Stephan, A. B., Kunz, H. H., Yang, E., & Schroeder, J. I. (2016). Rapid hyperosmotic-induced Ca_{2+} responses in Arabidopsis thaliana exhibit sensory potentiation and involvement of plastidial KEA transporters. *Proceedings of the National Academy of Sciences, 113*(42), E5242–E5249. https://doi.org/10.1073/pnas.1519555113.
47. Singh, A., et al. (2020). Monitoring soil salinity and recent advances in the mechanism of salinity tolerance in plants. *Biogeosystem Technique, 7*(2), 66–78. https://doi.org/10.13187/bgt.2020.2.66.
48. Feng, W., Kita, D., Peaucelle, A., Cartwright, H. N., Doan, V., Duan, Q., Liu, M.-C., et al. (2018). The FERONIA receptor kinase maintains cell wall integrity during salt stress through Ca_{2+} signaling. *Current Biology, 28*(5), 666–675.e5. https://doi.org/10.1016/j.cub.2018.01.023.
49. Pan, Y., Chai, X., Gao, Q., Zhou, L., Zhang, S., Li, L., & Luan, S. (2019). Dynamic interactions of plant CNGC subunits and calmodulins drive oscillatory Ca_{2+} channel

activities. *Developmental Cell, 48*(5), 710–725.e5. https://doi.org/10.1016/j.devcel.2018.12.025.

50. Tian, W., Hou, C., Ren, Z., Wang, C., Zhao, F., Dahlbeck, D., Hu, S., et al. (2019). A calmodulin-gated calcium channel links pathogen patterns to plant immunity. *Nature, 572*(7767), 131–135. https://doi.org/10.1038/s41586-019-1404-0.
51. Wu, F., Chi, Y., Jiang, Z., Xu, Y., Xie, L., Huang, F., Wan, D., et al. (2020). Hydrogen peroxide sensor HPCA1 is an LRR receptor kinase in Arabidopsis. *Nature, 578*(7794), 577–581. https://doi.org/10.1038/s41586-020-2032-3.
52. Rajput, V. D., Minkina, T., Feizi, M., Kumari, A., Khan, M., Mandzhieva, S., Sushkova, S., El-Ramady, H., Verma, K. K., Singh, A., Hullebusch, E. D. van, Singh, R. K., Jatav, H. S., & Choudhary, R. (2021). Effects of silicon and silicon-based nanoparticles on rhizosphere microbiome, plant stress and growth. *Biology, 10*(8), 791. https://doi.org/10.3390/biology10080791.
53. Quan, R., Lin, H., Mendoza, I., Zhang, Y., Cao, W., Yang, Y., Shang, M., et al. (2007). SCABP8/CBL10, a putative calcium sensor, interacts with the protein kinase SOS_2 to protect Arabidopsis shoots from salt stress. *The Plant Cell, 19*(5), 1415–1431. https://doi.org/10.1105/tpc.106.042291.
54. Rajput, V. D., Minkina, T., Kumari, A., Shende, S. S., Ranjan, A., Faizan, M., Barakhov, A., Gromovik, A., Gorbunova, N., Rajput, P., Singh, A., Khabirov, I., Nazarenko, O., Sushkova, S., & Kizilkaya, R. (2022). A review on nanobioremediation approaches for restoration of contaminated soil. *Eurasian Journal of Soil Science (EJSS), 11*(1), 43–60. https://doi.org/10.18393/ejss.990605.
55. Nieves-Cordones, M., Alemán, F., Martínez, V., & Rubio, F. (2014). K+ uptake in plant roots. The systems involved, their regulation, and parallels in other organisms. *Journal of Plant Physiology, 171*(8), 688–695. https://doi.org/10.1016/j.jplph.2013.09.021.
56. Xu, J., Li, H.-D., Chen, L.-Q., Wang, Y., Liu, L.-L., He, L., & Wu, W.-H. (2006). A protein kinase, interacting with two calcineurin B-like proteins, regulates K+ transporter AKT1 in Arabidopsis. *Cell, 125*(7), 1347–1360. https://doi.org/10.1016/j.cell.2006.06.011.
57. Sánchez-Barrena, M. J., Chaves-Sanjuan, A., Raddatz, N., Mendoza, I., Cortés, Á., Gago, F., González-Rubio, J. M., et al. (2020). Recognition and activation of the plant AKT1 potassium channel by the kinase $CIPK_{23}$. *Plant Physiology, 182*(4), 2143–2153. https://doi.org/10.1104/pp.19.01084.
58. Cheong, Y. H., Pandey, G. K., Grant, J. J., Batistic, O., Li, L., Kim, B.-G., Lee, S.-C., et al. (2007). Two calcineurin B-like calcium sensors, interacting with protein kinase $CIPK_{23}$, regulate leaf transpiration and root potassium uptake in Arabidopsis. *The Plant Journal, 52*(2), 223–239. https://doi.org/10.1111/j.1365-313X.2007.03236.x.
59. Ren, X.-L., Qi, G.-N., Feng, H.-Q., Zhao, S., Zhao, S.-S., Wang, Y., & Wu, W.-H. (2013). Calcineurin B-like protein CBL_{10} directly interacts with AKT_1 and modulates K+ homeostasis in Arabidopsis. *Plant Journal, 74*(2), 258–266. https://doi.org/10.1111/tpj.12123.
60. Covington, M. F., Maloof, J. N., Straume, M., Kay, S. A., & Harmer, S. L. (2008). Global transcriptome analysis reveals circadian regulation of key pathways in plant growth and development. *Genome Biology, 9*(8), R130. https://doi.org/10.1186/gb-2008-9-8-r130.

61. Wei, H., Wang, X., He, Y., Xu, H., & Wang, L. (2021). Clock component OsPRR73 positively regulates rice salt tolerance by modulating OsHKT2;1-mediated sodium homeostasis. *EMBO Journal, 40,* e105086. https://doi.org/10.15252/embj.2020105086.
62. Ayadi, M., Brini, F., & Masmoudi, K. (2019). Overexpression of a wheat aquaporin gene, TdPIP2;1, enhances salt and drought tolerance in transgenic durum wheat cv. Maali. *International Journal of Molecular Sciences, 20*(10), 2389. https://doi.org/10.3390/ijms20102389.
63. Yang, Y., & Guo, Y. (2017). Elucidating the molecular mechanisms mediating plant salt-stress responses. *New Phytologist, 217*(2), 523–539. https://doi.org/10.1111/nph.14920.
64. Marusig, D., & Tombesi, S. (2020). Abscisic acid mediates drought and salt stress responses in Vitis vinifera—A review. *International Journal of Molecular Sciences, 21*(22), 8648. https://doi.org/10.3390/ijms21228648.
65. Zhou, X., Naguro, I., Ichijo, H., & Watanabe, K. (2016). Mitogen-activated protein kinases as key players in osmotic stress signaling. *Biochimica et Biophysica Acta – General Subjects, 1860*(9), 2037–2052. https://doi.org/10.1016/j.bbagen.2016.05.032.
66. Pan, J., Zhang, M., Kong, X., Xing, X., Liu, Y., Zhou, Y., Liu, Y., et al. (2011). ZmMPK17, a novel maize group D MAP kinase gene, is involved in multiple stress responses. *Planta, 235*(3), 661–676. https://doi.org/10.1007/s00425-011-1510-0.
67. Yu, Z., Duan, X., Luo, L., Dai, S., Ding, Z., & Xia, G. (2007). Two rice cytosolic ascorbate peroxidases differentially improve salt tolerance in transgenic Arabidopsis. *Plant Cell Reports, 26*(11), 1909–1917. https://doi.org/10.1007/s00299-007-0395-7.
68. Sharma, R., Sahoo, A., Devendran, R., & Jain, M. (2014). Over-expression of a rice tau class glutathione S-transferase gene improves tolerance to salinity and oxidative stresses in Arabidopsis. *PLoS ONE, 9*(3), e92900. https://doi.org/10.1371/journal.pone.0092900.
69. Yu, Z., Duan, X., Luo, L., Dai, S., Ding, Z., & Xia, G. (2020). How plant hormones mediate salt stress responses. *Trends in Plant Science, 25*(12), 1117–1130. https://doi.org/10.1016/j.tplants.2020.06.008.
70. Verma, V., Ravindran, P., & Kumar, P. P. (2016). Plant hormone-mediated regulation of stress responses. *BMC Plant Biology, 16,* 1–10. https://doi.org/10.1186/s12870-016-0771-y.
71. Zhao, Y., Zhang, Z., Gao, J., Wang, P., Hu, T., Wang, Z., Hou, Y.-J., et al. (2018). Arabidopsis duodecuple mutant of PYL ABA receptors reveals PYL repression of ABA-independent SnRK2 activity. *Cell Reports, 23*(11), 3340–3351.e5. https://doi.org/10.1016/j.celrep.2018.05.044.
72. Fujita, Y., Yoshida, T., & Yamaguchi-Shinozaki, K. (2013). Pivotal role of the AREB/ABF-SnRK2 pathway in ABRE-mediated transcription in response to osmotic stress in plants. *Physiologia Plantarum, 147*(1), 15–27. https://doi.org/10.1111/j.1399-3054.2012.01635.x.
73. Su, Q., Zheng, X., Tian, Y., & Wang, C. (2020). Exogenous brassinolide alleviates salt stress in *Malus hupehensis* Rehd. by regulating the transcription of NHX-type Na+(K+)/H+ antiporters. *Frontiers in Plant Science, 11*, 38. https://doi.org/10.3389/fpls.2020.00038.

74. Rajput, P., Rajput, V. D., Singh, A., Sharma, R., Rawat, S., Tomar, R. S., Singh, A., & Minkina, T. (2021). Impact of emerging tools for solving 'hidden hunger' in humans. *Emerging Tools for Sustainable Agriculture and Food Security*, *1*(1), 17–30.
75. Lian, N., Wang, X., Jing, Y., & Lin, J. (2021). Regulation of cytoskeleton-associated protein activities: Linking cellular signals to plant cytoskeletal function. *Journal of Integrative Plant Biology, 63*(2), 241–250. https://doi.org/10.1111/jipb.13046.
76. Ma, D., & Han, R. (2020). Microtubule organization defects in *Arabidopsis thaliana*. *Plant Biology, 22*(6), 971–980. https://doi.org/10.1111/plb.13114.
77. Liu, S. G., Zhu, D. Z., Chen, G. H., Gao, X.-Q., & Zhang, X. S. (2012). Disrupted actin dynamics trigger an increment in the reactive oxygen species levels in the Arabidopsis root under salt stress. *Plant Cell Reports, 31*(7), 1219–1226. https://doi.org/10.1007/s00299-012-1242-z.
78. Qian, D., & Xiang, Y. (2019). Actin cytoskeleton as an actor in upstream and downstream of calcium signaling in plant cells. *International Journal of Molecular Sciences, 20*(6), 1403. https://doi.org/10.3390/ijms20061403.
79. Singh, A., Rajput, V. D., Rawat, S., Sharma, R., Kumar Singh, A., Kumar Singh, A., & Tomar, R. S. (2021). Emerging tools for sustainable agriculture and food security: An introduction. *Emerging Tools for Sustainable Agriculture and Food Security, 1*(1), 3–16.
80. Monniaux, M., & Hay, A. (2016). Cells, walls, and endless forms. *Current Opinion in Plant Biology, 34*, 114–121. https://doi.org/10.1016/j.pbi.2016.10.010.
81. Zhao, C., Zayed, O., Yu, Z., Jiang, W., Zhu, P., Hsu, C.-C., Zhang, L., Tao, W. A., Lozano-Durán, R., & Zhu, J.-K. (2018). Leucine-rich repeat extensin proteins regulate plant salt tolerance in Arabidopsis. *Proceedings of the National Academy of Sciences of the United States of America, 115*(49), 13123–13128. https://doi.org/10.1073/pnas.1816991115.
82. Du, Q., & Wang, H. (2015). The role of HD-ZIP III transcription factors and miR165/166 in vascular development and secondary cell wall formation. *Plant Signaling & Behavior, 10*(9), e1078955. https://doi.org/10.1080/15592324.2015.1078955.
83. Endler, A., & Persson, S. (2011). Cellulose synthases and synthesis in Arabidopsis. *Molecular Plant, 4*, 199–211. https://doi.org/10.1093/mp/ssq079.
84. Liu, C. Y., Rao, X. L., Li, L. G., & Dixon, R. A. (2021). Abscisic acid regulates secondary cell wall formation and lignin deposition in Arabidopsis thaliana through phosphorylation of NST1. *Proceedings of the National Academy of Sciences of the United States of America, 118*(15), e2010911118. https://doi.org/10.1073/pnas.2010911118.
85. Machado, R. M. A., & Serralheiro, R. P. (2017). Soil salinity: Effect on vegetable crop growth. Management practices to prevent and mitigate soil salinization. *Horticulturae, 3*(2), 30. https://doi.org/10.3390/horticulturae3020030.
86. Hu, Y., & Schmidhalter, U. (2002). Limitation of salt stress to plant growth. In B. Hock & C. F. Elstner (Eds.), *Plant Toxicology* (pp. 91–224). Marcel Dekker Inc.
87. Bano, A., & Fatima, M. (2009). Salt tolerance in *Zea mays* (L.) following inoculation with Rhizobium and Pseudomonas. *Biology and Fertility of Soils, 45*, 405–413. https://doi.org/10.1007/s00374-008-0338-y.

88. Netondo, G. W., Onyango, J. C., & Beck, E. (2004). Sorghum and salinity: II. Gas exchange and chlorophyll fluorescence of sorghum under salt stress. *Crop Science, 44*(3), 806–811. https://doi.org/10.2135/cropsci2004.8060.
89. Ashraf, M. (2004). Some important physiological selection criteria for salt tolerance in plants. *Flora-Morphology, Distribution, Functional Ecology of Plants, 199*(5), 361–376.
90. Seckin, B., Sekmen, A. H., & Turkan, I. (2009). An enhancing effect of exogenous mannitol on the antioxidant enzyme activities in roots of wheat under salt stress. *Journal of Plant Growth Regulation, 28*, 12–20. https://doi.org/10.1007/s00344-008-9076-8.
91. Wichern, J., Wichern, F., & Joergensen, R. G. (2006). Impact of salinity on soil microbial communities and the decomposition of maize in acidic soils. *Geoderma, 137*, 100–108. https://doi.org/10.1016/j.geoderma.2006.08.004.
92. Andronov, E. E., Petrova, S. N., Pinaev, A. G., Pershina, E. V., Rakhimgalieva, S. Z., Akhmeden, K. M., & Sergaliev, N. K. (2012). Analysis of the structure of the microbial community in soils with different degrees of salinization using T-RFLP and real-time PCR techniques. *Eurasian Soil Science, 45*, 147–156.
93. Xu, X., Shi, Z., Li, D., Rey, A., Ruan, H., Craine, J. M., Liang, J., Zhou, J., & Luo, Y. (2016). Soil properties control decomposition of soil organic carbon: Results from data-assimilation analysis. *Geoderma, 262*, 235–242. https://doi.org/10.1016/j.geoderma.2015.08.027.
94. Cossins, D. (2014). Next generation: Nanoparticles augment plant functions. The incorporation of synthetic nanoparticles into plants can enhance photosynthesis and transform leaves into biochemical sensors. *The Scientist*. Retrieved from: https://www.the-scientist.com/ (accessed on 20 June 2024).
95. Siddiqui, M. H., & Al-Whaibi, M. H. (2014). Role of nano-SiO_2 in the germination of tomato (*Lycopersicum esculentum* seeds Mill.). *Saudi Journal of Biological Sciences, 21*, 13–17. https://doi.org/10.1016/j.sjbs.2013.04.005.
96. Khodakovskaya, M. V., de Silva, K., Biris, A. S., Dervishi, E., & Villagarcia, H. (2012). Carbon nanotubes induce growth enhancement of tobacco cells. *ACS Nano, 6*(3), 2128–2135. https://doi.org/10.1021/nn204643k.
97. Farooq, M. A., & Dietz, K. J. (2015). Silicon as versatile player in plant and human biology: Overlooked and poorly understood. *Frontiers in Plant Science, 6*, 1–14. https://doi.org/10.3389/fpls.2015.00124.
98. Alsaeedi, A. H., El-Ramady, H., Alshaal, T., El-Garawani, M., Elhawat, N., & Almohsen, M. (2017). Engineered silica nanoparticles alleviate the detrimental effects of Na+ stress on germination and growth of common bean (*Phaseolus vulgaris*). *Environmental Science and Pollution Research, 24*, 21917–21928. https://doi.org/10.1007/s11356-017-9882-6.
99. Ma, X., Wang, Q., Rossi, L., Ebbs, S. D., & White, J. C. (2016). Multigenerational exposure to cerium oxide nanoparticles: Physiological and biochemical analysis reveals transmissible changes in rapid cycling *Brassica rapa*. *NanoImpact, 1*, 46–54. https://doi.org/10.1016/j.impact.2016.05.001.
100. Rossi, L., Zhang, W., & Ma, X. (2017). Cerium oxide nanoparticles alter the salt stress tolerance of *Brassica napus* L. by modifying the formation of root apoplastic

barriers. *Environmental Pollution, 229*, 132–138. https://doi.org/10.1016/j.envpol.2017.05.065.

101. Singh, P., & Nanda, A. (2014). Enhanced sun protection of nano-sized metal oxide particles over conventional metal oxide particles: An in vitro comparative study. *International Journal of Cosmetic Science, 36*(3), 273–283. https://doi.org/10.1111/ics.12117.
102. Ramesh, M., Palanisamy, K., Babu, K., & Sharma, N. K. (2014). Effects of bulk and nano-titanium dioxide and zinc oxide on physio-morphological changes in *Triticum aestivum* Linn. *Journal of Global Biosciences, 3*(1), 415–422.
103. Oliveira, H. C., Gomes, B. C. R., Pelegrino, M. T., & Seabra, A. B. (2016). Nitric oxide-releasing chitosan nanoparticles alleviate the effects of salt stress in maize plants. *Nitric Oxide, 61*, 10–19. https://doi.org/10.1016/j.niox.2016.10.001.
104. Farnese, F. S., Menezes-Silva, P. E., Gusman, G. S., & Oliveira, J. A. (2016). When bad guys become good ones: The key role of reactive oxygen species and nitric oxide in the plant responses to abiotic stress. *Frontiers in Plant Science, 7*, 471. https://doi.org/10.3389/fpls.2016.00471.
105. Zulfiqar, F., & Ashraf, M. (2021). Nanoparticles potentially mediate salt stress tolerance in plants. *Plant Physiology and Biochemistry, 160*, 257–268. https://doi.org/10.1016/j.plaphy.2021.02.032.
106. Tortella, G. R., Rubilar, O., Durán, N., Diez, M. C., Martínez, M., Parada, J., & Seabra, A. B. (2020). Silver nanoparticles: Toxicity in model organisms as an overview of its hazard for human health and the environment. *Journal of Hazardous Materials, 390*, Article 121974. https://doi.org/10.1016/j.jhazmat.2019.121974.
107. Hernández-Hernández, H., Juárez-Maldonado, A., Benavides-Mendoza, A., Ortega-Ortiz, H., Cadenas-Pliego, G., Sánchez-Aspeytia, D., & González-Morales, S. (2018). Chitosan-PVA and copper nanoparticles improve growth and overexpress the SOD and JA genes in tomato plants under salt stress. *Agronomy, 8*, 175. https://doi.org/10.3390/agronomy8090175.
108. Shahi, S., & Srivastava, M. (2018). Influence of foliar application of manganese on growth, pigment content, and nitrate reductase activity of *Vigna radiata* (L.) R. Wilczek under salinity. *Journal of Plant Nutrition, 41*(10), 1397–1408. https://doi.org/10.1080/01904167.2018.1444465.
109. Ali, S., Mehmood, A., & Khan, N. (2021). Uptake, translocation, and consequences of nanomaterials on plant growth and stress adaptation. *Journal of Nanomaterials, 2021*, Article 5559183. https://doi.org/10.1155/2021/5559183.
110. Dev, A., Srivastava, A. K., & Karmakar, S. (2018). Nanomaterial toxicity for plants. *Environmental Chemistry Letters, 16*(1), 85–100. https://doi.org/10.1007/s10311-017-0647-y.
111. Zulfiqar, U., Subhani, T., & Husain, S. W. (2016). Synthesis of silica nanoparticles from sodium silicate under alkaline conditions. *Journal of Sol-Gel Science and Technology, 77*, 753–758. https://doi.org/10.1007/s10971-015-3950-7.
112. Ali, S., Rizwan, M., Qayyum, M. F., Ok, Y. S., Ibrahim, M., Riaz, M., & Shahzad, A. N. (2017). Biochar soil amendment on alleviation of drought and salt stress in plants: A critical review. *Environmental Science and Pollution Research International, 24*(14), 12700–12712. https://doi.org/10.1007/s11356-017-8904-x.

113. Zhang, Y., Ding, J., Wang, H., Su, L., & Zhao, C. (2020). Biochar addition alleviates the negative effects of drought and salinity stress on soybean productivity and water use efficiency. *BMC Plant Biology, 20*, Article 1–11. https://doi.org/10.1186/s12870-019-2220-6.
114. Ghassemi-Golezani, K., & Farhangi-Abriz, S. (2021). Biochar-based metal oxide nanocomposites of magnesium and manganese improved root development and productivity of safflower (*Carthamus tinctorius* L.) under salt stress. *Rhizosphere, 19*, Article 100416. https://doi.org/10.1016/j.rhisph.2021.100416.
115. Ünlükara, A., Kurunc, A., Kesmez, G. D., Yurtseven, E., & Suarez, D. L. (2010). Effects of salinity on eggplant (*Solanum melongena* L.) growth and evapotranspiration. *Irrigation and Drainage: The Journal of the International Commission on Irrigation and Drainage, 59*(2), 203–214. https://doi.org/10.1002/ird.528.
116. Chourasia, K. N., Lal, M. K., Tiwari, R. K., Dev, D., Kardile, H. B., Patil, V. U., & Bisht, J. K. (2021). Salinity stress in potato: Understanding physiological, biochemical and molecular responses. *Life, 11*(6), Article 545. https://doi.org/10.3390/life11060545.
117. Temel, A., & Gozukirmizi, N. (2015). Physiological and molecular changes in barley and wheat under salinity. *Applied Biochemistry and Biotechnology, 175*(6), 2950–2960. https://doi.org/10.1007/s12010-014-1464-3.
118. Faizan, M., Bhat, J. A., Chen, C., Alyemeni, M. N., Wijaya, L., Ahmad, P., & Yu, F. (2021). Zinc oxide nanoparticles (ZnO-NPs) induce salt tolerance by improving the antioxidant system and photosynthetic machinery in tomatoes. *Plant Physiology and Biochemistry, 161*, 122–130. https://doi.org/10.1016/j.plaphy.2021.03.005.
119. Wang, Z., Li, H., Li, X., Xin, C., Si, J., Li, S., & Wang, F. (2020). Nano-ZnO priming induces salt tolerance by promoting photosynthetic carbon assimilation in wheat. *Archives of Agronomy and Soil Science, 66*(8), 1259–1273. https://doi.org/10.1080/03650340.2020.1755180.
120. Gohari, G., Safai, F., Panahirad, S., Akbari, A., Dadpour, M. R., & Fotopoulos, V. (2020). Modified multiwall carbon nanotubes display either phytotoxic or growth-promoting and stress-protecting activity in *Ocimum basilicum* L. in a concentration-dependent manner. *Chemosphere, 249*, Article 126171. https://doi.org/10.1016/j.chemosphere.2020.126171.
121. Pinedo-Guerrero, Z. H., Cadenas-Pliego, G., Ortega-Ortiz, H., González-Morales, S., Benavides-Mendoza, A., Valdés-Reyna, J., & Juárez-Maldonado, A. (2020). Form of silica improves yield, fruit quality, and the antioxidant defense system of tomato plants under salt stress. *Agriculture, 10*, 377. https://doi.org/10.3390/agriculture10080377.
122. Mahmoud, L. M., Dutt, M., Shalan, A. M., El-Kady, M. E., El-Boray, M. S., Shabana, Y. M., & Grosser, J. W. (2020). Silicon nanoparticles mitigate oxidative stress of in vitro-derived banana (*Musa acuminata* 'Grand Nain') under simulated water deficit or salinity stress. *South African Journal of Botany, 132*, 155–163. https://doi.org/10.1016/j.sajb.2020.04.028.
123. Rajput, V. D., Minkina, T., Kumari, A., Harish, Singh, V. K., Verma, K. K., Mandzhieva, S., Sushkova, S., Srivastava, S., & Keswani, C. (2021). Coping with the challenges of abiotic stress in plants: New dimensions in the field application of nanoparticles. *Plants, 10*(6), 1221. https://doi.org/10.3390/plants10061221.
124. Rajput, V. D., Minkina, T., Sushkova, S., Behal, A., Maksimov, A., Blicharska, E., Ghazaryan, K., & Movsesyan, H. (2019). ZnO and CuO nanoparticles: A threat to

soil organisms, plants, and human health. *Environmental Geochemistry and Health, 42*(1), 147–158. https://doi.org/10.1007/s10653-019-00317-3.

125. Rajput, V. D., Minkina, T., Sushkova, S., Tsitsuashvili, V., Mandzhieva, S., Gorovtsov, A., Nevidomskyaya, D., & Gromakova, N. (2017). Effect of nanoparticles on crops and soil microbial communities. *Journal of Soils and Sediments, 18*(6), 2179–2187. https://doi.org/10.1007/s11368-017-1793-2.
126. Rajput, V. D., Fedorenko, A., Minkina, T., Sushkova, S., Mandzhieva, S., Lysenko, V., Duplii, N., Fedorenko, G., Dvadnenko, K., & Ghazaryan, K. (2018). Toxicity of copper oxide nanoparticles on spring barley (*Hordeum sativum* distichum). *Science of The Total Environment, 645*, 1103–1113. https://doi.org/10.1016/j.scitotenv.2018.07.211.
127. Rajput, V. D., Kumari, A., Minkina, T., Barakhov, A., Singh, S., Mandzhieva, S. S., Sushkova, S., Ranjan, A., Rajput, P., & Garg, M. C. (2022). A practical evaluation of the integrated role of biochar and nanomaterials in soil remediation processes. *Environmental Geochemistry and Health, 45, 9435–9449.* https://doi.org/10.1007/s10653-022-01375-w.
128. Rajput, V. D., Faizan, M., Upadhyay, S. K., Kumari, A., Ranjan, A., Sushkova, S., Chauhan, P. K., Mahmud, A. A., Burachevskaya, M., Chaplygin, V., & Deryabkina, I. (2022). Influence of nanoparticles on the plant rhizosphere microbiome. *Sustainable Plant Nutrition in a Changing World*, 83–102. https://doi.org/10.1007/978-3-030-97389-6_4.
129. Rajput, V. D., Singh, A., Minkina, T., Rawat, S., Mandzhieva, S., Sushkova, S., Shuvaeva, V., Nazarenko, O., Rajput, P., Komariah, Verma, K. K., Singh, A. K., Rao, M., & Upadhyay, S. K. (2021). Nano-enabled products: Challenges and opportunities for sustainable agriculture. *Plants, 10*(12), 2727. https://doi.org/10.3390/plants10122727.
130. Singh, A., Singh, A. K., Rawat, S., Pal, N., Rajput, V. D., Minkina, T., Sharma, R., Singh, N. P., & Tripathi, J. N. (2022). Satellite-based quantification of methane emissions from wetlands and rice paddies ecosystems in North and Northeast India. *Hydrobiology, 1*(3), 317–330. https://doi.org/10.3390/hydrobiology1030023.
131. Singh, A., Sengar, R. S., Rajput, V. D., Minkina, T., & Singh, R. K. (2022). Zinc oxide nanoparticles improve salt tolerance in rice seedlings by improving physiological and biochemical indices. *Agriculture, 12*(7), 1014. https://doi.org/10.3390/agriculture12071014.
132. Singh, A., Sengar, R. S., Sharma, R., Rajput, P. (2021). Nano-priming technology for sustainable agriculture. *Biogeosystem Technique, 8*(2), 79–91. https://doi.org/10.13187/bgt.2021.2.79.
133. Singh, A., Mehrotra, R., Rajput, V. D., Dmitriev, P., Singh, A. K., Kumar, P., Tomar, R. S., Singh, O., & Singh, A. K. (2022). Geoinformatics, artificial intelligence, sensor technology, big data. *Sustainable Agriculture Systems and Technologies*, 295–313. https://doi.org/10.1002/9781119808565.ch14.
134. Singh, A., Rajput, V. D., Rawat, S., Sharma, R., Singh, A. K., Kumar, P., Singh, A. K., Minkina, T., Singh, R. P., & Singh, S. (2022). Geoinformatics and nanotechnological approaches for coping up abiotic and biotic stress in crop plants. *Sustainable Agriculture Systems and Technologies*, 337–359. https://doi.org/10.1002/9781119808565.ch17.

CHAPTER 6

Application of Nanoparticles and Nano-Biochar for Improving Soil Physiochemical Properties

SHEETAL KUMARI and MANOJ CHANDRA GARG

Amity Institute of Environmental Science (AIES), Amity University, Noida, Uttar Pradesh, India

ABSTRACT

Global climate change has many negative consequences for all life forms. Outstanding examples include salinity and drought, which are regarded as multidimensional stresses that have a negative impact on plant growth and sustainable agriculture. Salt stress is a significant threat to agriculture in the age of climate change, reducing crop productivity by imposing primary and secondary effects. Advanced nanostructured materials, such as nano biochar, have developed long-term solutions to a variety of new challenges. Carbon nanomaterials have emerged as strong tools in recent years due to their unique properties and wide range of uses in fields such as energy, agriculture, materials, and the environment. Nanobiochar has significant potential, particularly for the remediation of toxic pollutants and the enhancement of agricultural yields. Nano biochar is a biochar substance and has improved physical, chemical, and surface properties. It has a variety of benefits, including improved plant growth and soil characteristics. Biochar is a viscous organic substance formed by the pyrolysis of plant or animal biomass. Because of its economic effectiveness, environmental friendliness, and sustainability, it may become the ideal alternative to traditional approaches. It can also help to prevent climate

Nano-Biochar and Nanoparticles for Soil Health Improvement.
Vishnu D. Rajput and Tatiana M. Minkina (Eds.)

change through its carbon sequestration capability. With the progress of nanotechnology, studies have been undertaken on the sustainable generation of nanobiochar for soil and agricultural applications.

6.1 INTRODUCTION

Biochar is now produced on a large scale all over the world and has recently gained popularity as a soil amendment. The thermochemical conversion of biomasses with no oxygen availability results in biochar [1]. Char is a prominent byproduct of biomass thermochemical reactions. Char is also known as biochar, especially when it is employed as a soil additive to enhance soil health. Furthermore, as biomass has a low sulfur and nitrogen content, burning it produces fewer hazardous gas emissions than burning most fossil fuels, including nitrous oxides (NOx) and sulfur dioxide (SO_2). Biomass is a promising source of renewable energy because of its benefits [2]. Thermochemical techniques like gasification and pyrolysis can transform biomass into biofuels and other bioproducts. The net carbon dioxide emissions from the use of biofuels are believed to be zero or negative because the emitted CO_2 was recycled from the atmosphere and gathered during photosynthesis [2]. The major byproducts of biomass thermochemical processes include syngas, bio-oil, charcoal, and tar, with yields varying according to process variables. Two significant intermediate products that can be utilized to produce fuels that can substitute for traditional fuels are syngas and bio-oil [3, 4]. Researchers are interested in using biochar to improve the environment since it is produced in an economical, sustainable, and recyclable approach [5, 6]. The use of biochar has demonstrated impressive benefits in reducing the harmful effects of saline stress and drought on plants, in addition to increasing the organic carbon content of the soil [7]. The potential of biochar to affect and change the biological, chemical, and physical characteristics of the soil [8]. The assessment of soil pH and the stability of soil aggregates serve as indicators of the physicochemical and biological characteristics of the soil. Under various conditions, the use of biochar in soil is controlled by its important functional indicators, such as crop yield, ecotoxicity, and erosion rate. Tsang et al. and his team described the efficient and goal-based techniques for biochar amendment. Many physicochemical features of biochar are also explored. Biochar is physically and chemically altered by using various raw materials and pyrolysis temperatures, which alter its pore distribution, cation exchange capacity, and surface functional groups on the adsorption of heavy metals [9]. However, to improve soil health and wastewater management, it is critical to assess and define the physicochemical features of biochar before applying it. The use of biochar in soil repair and soil quality maintenance is critically examined in this chapter [10]. Applications based on biochar are relatively new and can address pollution and climate change.

6.1.1 ORIGIN OF BIOCHAR

The "Terra-Preta" soils, often known as the Amazonian region's soils, are related to the origin of biochar. Due to their vastly increased crop productivity compared to the nearby sterile tropical soils, these soils have received attention from around the world [11]. Additional thorough investigation indicated that these soils are assumed to have utilized biochar as a crucial component, which helps to explain some of the peculiar characteristics of Terra-Preta soil [12]. There are different types of methods for the preparation of biochar as shown in Figure 6.1. A term newly coined to describe renewable fuel, soil improvement, and carbon sequestration is 'biochar.'

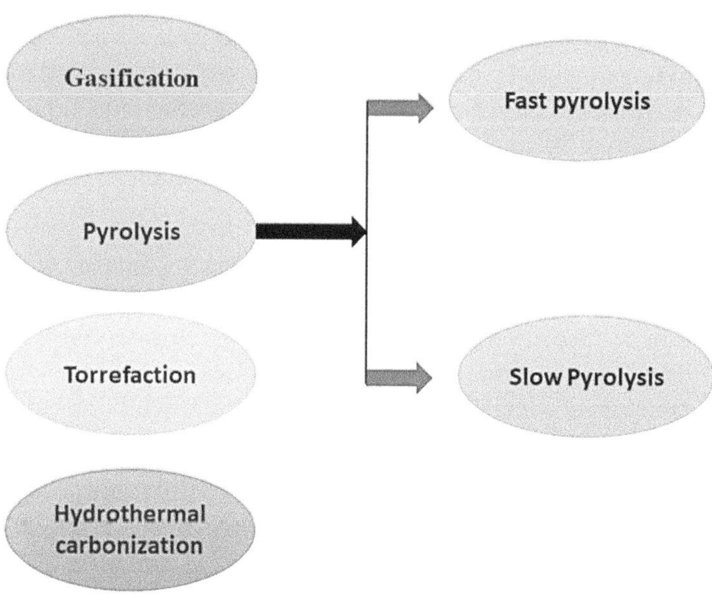

FIGURE 6.1 Methods for the preparation of biochar.

6.2 PRODUCTION OF NANO-BIOCHAR

A solid carbonaceous product called biochar is made from biomass feedstock, which includes agricultural waste and other lignocellulosic materials, through a controlled thermal decomposition process that occurs by heating at a temperature between 350 and 700°C while taking place in an environment with little to no oxygen or oxygen in limited amounts (also known as pyrolysis or gasification) [13–16].

1. **Pyrolysis:** It is the process used most frequently to generate biochar. According to the rate of heating and residence duration, pyrolysis can be divided into two types: slow and fast. In slow pyrolysis, commonly referred to as traditional carbonization, biomass is heated at a low heating rate for a substantial amount of time to form biochar. Biochar is produced through quick pyrolysis with a high heating rate (about 200 K/min) and a brief residence time. The pyrolysis methods differ in that slow pyrolysis encourages high yields of biochar whereas quick pyrolysis encourages high yields of bio-oil.
2. **Gasification:** It is the process by which biomass is converted into a gaseous mixture by adding a regulated amount of an oxidizing agent at a high temperature that is largely made up of syngas (which comprises CO, H_2, CO_2, CH_4, and trace amounts of higher hydrocarbons) (greater than 700°C). About 10 wt.% of biomass is typically converted to biochar during gasification [17]. Syngas produced by air gasification has a low heating value of 4–7 MJ/Nm^3, whereas syngas produced by steam gasification has a high heating value of 10–14 MJ/Nm^3 [18].

   The several lignocelluloses (cellulose, hemicellulose, and lignin) breakdown processes and mechanisms are essential for the production of biochar from plant-based biomass during pyrolysis [19]. Thus, the specific carbonization breakdown behaviors of each component – cellulose, hemicellulose, and lignin depend on the temperature and heating rates (Table 6.1).
3. **Torrefaction:** Commonly referred to as mild pyrolysis, is the process of heating biomass to temperatures between 200 and 300°C for durations ranging from 30 minutes to several hours in an inert atmosphere [21]. As a crucial pre-processing step for enhancing the physicochemical qualities of biomass for combustion, the

TABLE 6.1 Mechanism of Lignocellulosic Biomass

Component	Temperature (°C)	Byproduct	References
Cellulose decomposition	315–400°C	Liquid product (tar or bio-oil)	[19, 20]
Hemicellulose decomposition	220–315°C	Solid product (biochar)	[20]
Lignin decomposition	160–900°C	Gaseous product (CH_4, CO, CO_2, H_2)	[20]

torrefaction process has drawn a lot of attention in the field of bioenergy. Only 10% of biomass's energy is lost as gases during this process, which leads to a 30% mass loss. Consequently, the torrefied solid product has a higher specific energy density [22].

4. **Hydrothermal Carbonization:** A thermochemical process known as wet torrefaction or hydrothermal carbonization converts organic feedstock into solid products with high carbon content. Friedrich Bergius initially suggested using the term "Hydrothermal Carbonization process" to refer to the natural coalification process in 1913. During HTC, biomass is heated in a restricted system under pressure (2–6 MPa) for 5–240 minutes while submerged in water at a temperature between 180 and 260°C [23]. The distinctive feature of the HTC process reduces the necessity to pre-dry wet biomass, which is a very energy-intensive procedure and a significant financial burden in biomass pre-processing, especially when carried out under high-temperature pre-treatments like slow pyrolysis and dry torrefaction [24]. Biochar yield and temperature from the different thermochemical processes are shown in Table 6.2.

TABLE 6.2 Biochar Yield from the Thermochemical Process

Process	Temperature	Biochar Yield (wt.%)	References
Fast pyrolysis	400–600	10–20	[11, 19, 20]
Slow pyrolysis	400–600	20–40	[12, 21]
Gasification	800–1,000	10	[16]
Hydrothermal carbonization	180–250	30–60	[22]

6.3 PROPERTIES OF BIOCHAR IN RELATION TO SOIL

Biochar is a flexible material for many environmental remediation techniques [25]. Biochar is particularly useful and important in soil remediation because it improves the properties of the soil and has a wide range of functional groups. These properties include superior porous assembly, surface charge, bulk density, high surface area, and water retention capacity [26, 27]. These characteristics of biochar can aid in the adsorption and immobilization of pollutants, which are accomplished by complexation, electrostatic interaction, and ion exchange [28].

1. **Surface Area:** The temperature at which the biochar was created as well as the kind of feedstock employed in its synthesis has an impact on the surface area of the biochar. The lowest limit of the range for the surface area of biochar, which ranges from 1.4 to 500 m_2 g_{-1}, was determined by Li et al. [29]. All studies showed that surface area grew as pyrolysis temperature increased, with the exception of those by Yue et al. and Angin, who asserted that surface area increased at lower temperatures and decreased as the temperature climbed [30, 31].
2. **Pore Characteristics:** Biochar is highly valuable because of its porous nature. By adsorbing plant nutrients, its micropores might increase soil fertility [32]. According to Liu et al., the porous characteristic of biochar may also help the soil retain water and offer protection for fungi and bacteria from protists and microarthropod predation [33].
3. **Soil Bulk Density:** Bulk density is a sign of healthy and compacted soil. It affects vital soil processes and production by influencing soil aeration, infiltration, water availability, plant nutrient availability, and soil microbial activity. Different interdependent characteristics of the soil, such as microbial population and plant growth, are significantly influenced by the water retention ability, bulk density, porosity, and redox conditions of the soil [34]. According to a study by Berihun and colleagues, adding biochar had a substantial impact on the porosity and bulk density of the soil [35]. Githinji et al. found that increasing the usage of biochar greatly decreased the bulk density as well [36]. According to Mukherjee et al., the application of biochar decreases the bulk weight of the soil because biochar has an incredibly high porosity and increases the pore capacity of

the soil, lowering the bulk density by a significant amount [37]. Additionally, it has been suggested that adding biochar to the soil over time may increase the amount of water and nutrients in the soil because it oxidizes over time after application and because biochar properties appear to be hydrophobic, increasing the soil's ability to absorb and hold onto water. By modifying soil porosity and aggregation level, the application of biochar successfully improved the soil's ability to retain water.

4. **Soil Aggregate Stability:** One of the most important factors that help soil sustain mechanical loads including the effects of rainfall, water erosion, and surface runoff is soil aggregate stability [38]. The soil structure, root growth, soil aeration, biota movement within the soil, and drought resistance are all enhanced by aggregates, which also protect organic materials.
5. **Hydraulic Conductivity:** Soil hydraulic conductivity may be affected by the application of biochar in one of three ways: up, down, or not at all. A maximum increase of 176% was noticed in clay soil that had been treated with biochar at a rate of 16 Mg ha_{-1} [39]. Based on the investigations, the effect of biochar on hydraulic conductivity can be summed up as follows: (i) In fine-textured soil (clay), significantly more striking was the rise in soil hydraulic conductivity; (ii) the decrease in soil hydraulic conductivity was more severe in soil with a coarse texture (sand); and (iii) the effect was minimal or non-existent in medium-textured soil.
6. **Soil Organic Matter (SOM):** The main supply of soil nutrients comes from soil organic matter (SOM). SOM can increase soil microbial activity, aid in nitrogen absorption, and lessen nitrogen losses through volatilization. Microbial biomass dissolved organic carbon, and light-fraction organic matter are all impacted by dynamic soil organic matter (DSOM), a vital part of the soil [40]. Biochar is naturally rich in organic carbon, significant metals, and aromatic compounds. By absorbing other nutrients, biochar can indirectly enhance the proportion of soil C/N ratio while also substantially improving soil humus and organic content [41]. Long-term benefits are anticipated from the soil advancements of biochar, while the use of biochar-ash composites as fertilizer promotes environmental sustainability and provides an immediate financial incentive [35].

6.4 APPLICATION OF BIOCHAR

The organic substance known as biochar has been approved for use in several agricultural practices for various purposes, including microbial growth support, disease management, pesticide remediation, fertilization, and improved plant development as mentioned in Figure 6.2 [42, 43]. Biochar was also found to increase crop growth, soil efficiency, and productivity [44, 45]. Some of its uses include soil remediation, CO_2 absorption, power storage, catalysis, redox reactions, and environmental protection [46, 47].

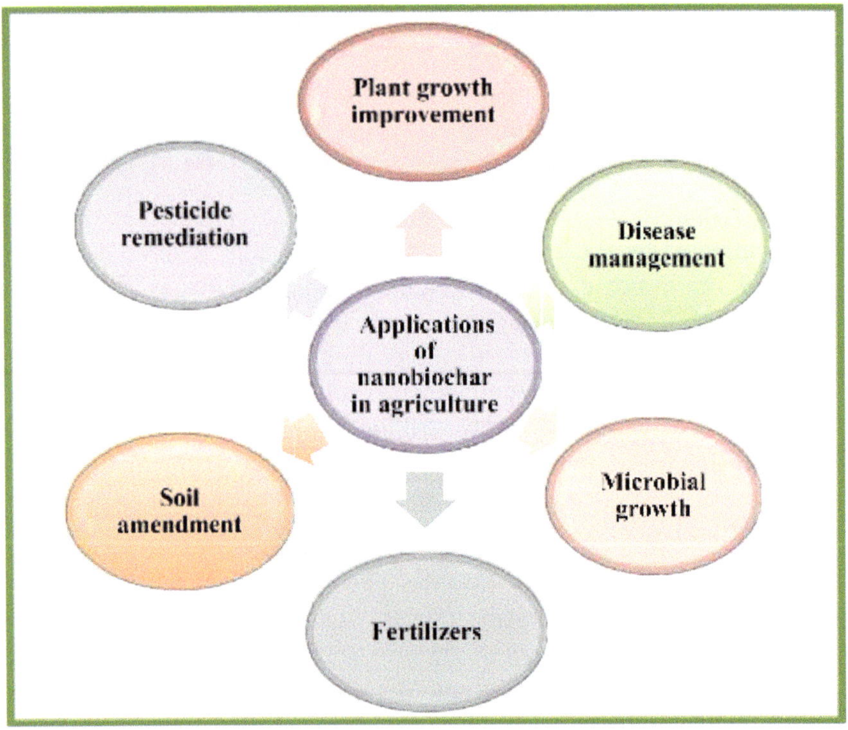

FIGURE 6.2 Application of nanobiochar in agriculture.

6.4.1 ENHANCING PLANT DEVELOPMENT AND AMENDING THE SOIL

Industrialization, deforestation, overgrazing, and other human activities have significantly decreased the number of nutrients in the soil. Lower yield and plant growth as a result of nutrient depletion have had a significant negative impact on food security [48]. Chemical fertilizers have been utilized excessively to maintain soil nutrients. In contrast, nutrient leakage and runoff reduce their bioavailability in soil and encourage the use of artificial fertilizers. This has caused soil contamination and its detrimental effects on soil biota in various areas [49]. Although a few options are being considered to address the issues of heavy nutrient application, scarcity of resources, and nutrient decline, there are few real-world alternatives [50]. In contrast to its use for carbon sequestration and environmental remediation, soil nutrient management with an emphasis on contaminated soils has received the majority of recent research regarding biochar [51]. In addition to improving soil fertility, permeability, and water-holding capacity, biochar also carries enormous amounts of nutrients due to its high charge density, resulting in an increase in crop output [42]. A crucial ingredient for plant growth is phosphorus (P). Phosphate anions (H_2PO_4) can be obtained by plants through phosphorous adsorption, desorption, and transition between soil particles and solution [52]. Therefore, phosphorous adsorption and desorption affect soil P, which influences phosphate uptake and usage by plants. Due to its ability to impact the soil, biochar can be used to amend soil because it has been shown to increase the availability of P and nitrogen and decrease nitrogen leaching, phosphorus fixation, and N_2O emissions [53].

Various investigations have suggested that using biochar in conjunction with specific chemicals, and physical and manufactured nanomaterials may have additional benefits like an increase in water retention, an increase in soil pH, improved cation exchange capacity, and an influence on plant nutrient uptake (Figure 6.3). A significant improvement in wheat growth was found when water-soluble carbon nanoparticles were combined with biochar [54]. It has higher soil pH, nutrient load, electrical conductivity (EC), and cation exchange capacity (CEC) because it contains organic material. Biochar may directly feed nutrients to the soil because it contains nutrients like N, P, and K. Additionally, by improving the soil's capacity to hold onto nutrients, biochar indirectly raises soil fertility. By limiting nutrient losses from the soil, biochar also lowers the potential for eutrophication in surface water bodies [51]. The increase in nutrient uptake by crops on soil amended with biochar was caused by an improvement in the soil nitrogen retention capacity in the Amazon region [55]. Using biochar as a soil amendment, according to Warnock et al.'s theory, "can change soil physicochemical properties, indirectly affecting mycorrhizae through influence on other soil microbes, change plant mycorrhizal fungi signaling processes, or detoxifies allelochemicals leading to altered root colonization by mycorrhizal fungi, and provide refugia from fungal grazers" [56]. According to studies by Yu et al., applying biochar to agricultural soils reduces the number of pesticides that plants take up from the soil, indicating that biochar has the capacity to hold onto organic chemicals [57].

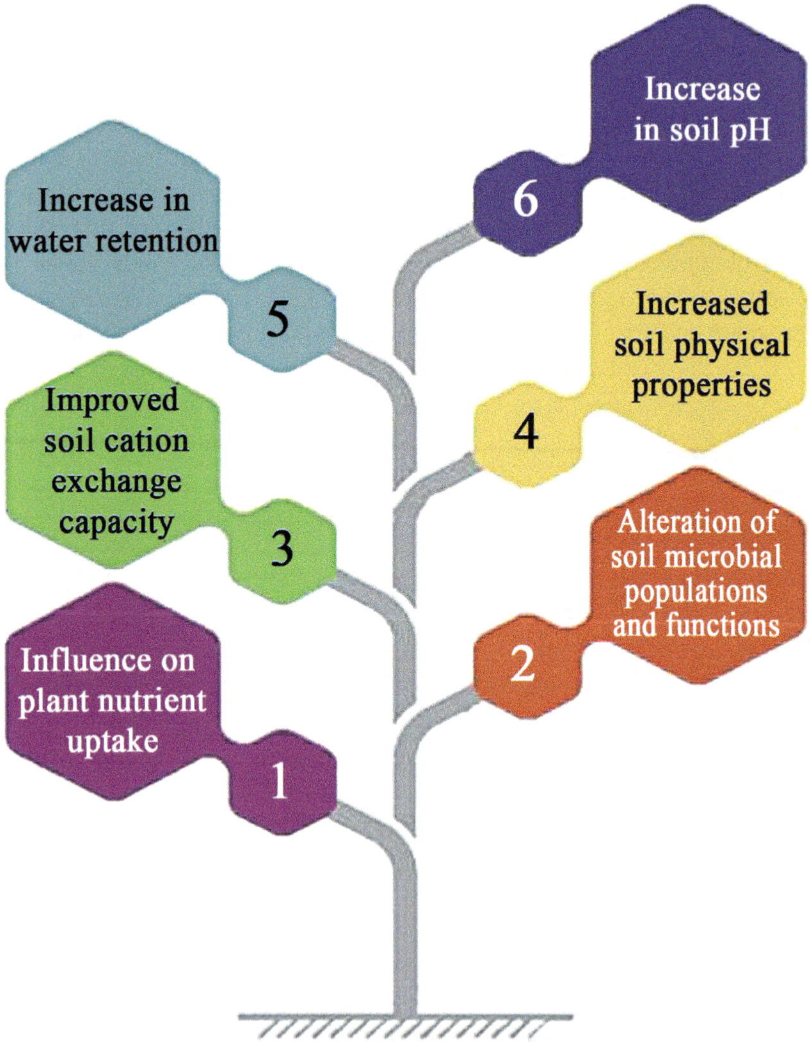

FIGURE 6.3 Mechanism of biochar as a soil conditioner for plant growth.

6.4.2 BIOCHAR-BASED GOOD/PREMIUM QUALITY FERTILIZER

Researchers have introduced biochar-based fertilizer to improve crop productivity and NUE since it has shown significant promise for usage as a slow-release fertilizer [58]. Due to its special physical and chemical characteristics and the addition of various types of nutrients to the biochar matrix, biochar can be employed as a vehicle for the delivery of nutrients. In comparison to traditional fertilizer, studies in the past have shown that fertilizer based on biochar could decrease nitrogen loss and boost the NUE by the crops over time. It is a high-performance organic fertilizer because biochar is made from biomass which contains higher concentrations of other micronutrients like copper, iron, manganese, zinc, and ash as well as macronutrients like magnesium, calcium, sulfur, and potassium [59]. The amount of total P and N in biochar produced from plant biomass, such as wood and green waste, was found to be higher than that produced from animal sources, such as broiler litter and sewage sludge [60]. Different nano biochar-based fertilizers and their effects are presented in Table 6.3. When compared to traditional fertilizers, biochar-based fertilizers exhibit various benefits. The following are some additional advantages over conventional fertilizer, in addition to improved slow-release performance:

- Increases water storage capacity [61];
- Potential to immobilize heavy metals [62];
- Utilized as a carbon–sequestering substance [63].

TABLE 6.3 Nano-biochar as a Fertilizer

Nano-Biochar Fertilizers (Crops/Plants)	Effects	References
Tomato, gram, rice	Crop production	[64–66]
Tomato	Seed germination	[14, 67]
Tobacco	Protein expressions	[64]
Wheat	Root elongation	[68]
Rice (MgO-modified biochar)	Increase phosphate retention	[69]
Gram	Enhanced water uptake	[15, 66]
Chinese cabbage	Growth and yield enhancement	[70]

6.4.3 BIOCHAR AS A BIOSORBENT FOR POLLUTANTS

Some pollutants in soil have been successfully absorbed by nano biochar (Table 6.4). By eliminating contaminants from soil and water, biochar can assist in reducing environmental problems [71]. Biochar's high surface area and microporosity cause organic pollutants in water to adsorb onto pores. As a result, biochar made at temperatures above 400°C and with a larger surface area is more effective at removing pollutants from water than biochar generated at temperatures lower than 400°C. In general, biochar is not a final carbonized product. The glassy domain exhibits an adsorption phase behavior for the carbonized fraction of biochar, while the rubbery domain exhibits a partition phase behavior for the non-carbonized portion. Partitioning of organic contaminants into the non-carbonized fraction is the main sorption process for biochar produced at low pyrolysis temperatures (less than 300°C), whereas adsorption onto porous carbonized fractions is the main sorption mechanism for biochar produced at high temperatures (higher than 400°C). This is due to the fact that the carbonized fraction of biochar increases as the pyrolysis temperature rises [72].

TABLE 6.4 Removal of Pollutants by Adsorption by Nano-Biochar

Biochar Feedstock	Nanocomposite Type	Pollutants Removed	Adsorption Capacity	References
Sugarcane bagasse	Sugarcane bagasse biochar	Methylene blue	6.2 mg/g	[73]
Pinewood	Pinewood nano biochar	Carbamazepine	95%	[74]
Hickory chips, bamboo bagasse	Clay-biochar composite	Methylene blue	7.90 mg/g	[75]
Milled cotton	Graphene (functional nanoparticle coated)	Methylene blue	174 mg/g	[76]
Pinewood	Manganese-oxide/ biochar composite (Mn/ BC)	Pb (II)	98.90%	[77]
Maize straw	Nano cerium oxide-maize straw biochar)	PO_{43-}	78 mg/g	[78]
Corn straw	Iron (Fe)-impregnated biochar (FBC)	As(V)	6.80 mg/g	[79]
Rice straw	Rice straw nano biochar	Antibiotic resistance genes-amp C, erm B	Adsorption of eDNA-nano 700 (60%)	[80]
Wheat straw	Magnetic nano biochar	Tetracycline and Hg (II)	268.3 mg/g, 127.4 mg/g,	[81]
Corn stover	Nano-ZnS/ZnO modified biochar	Cu (II)	924.5 mg/g	[82]

6.4.4 BIOCHAR AS A CATALYST

6.4.4.1 CATALYST FOR SYNGAS

Gasification of biomass results in the production of syngas, which contain a significant amount of tar that is harmful to subsequent operations. There are several ways to get rid of the tars in syngas, including water or oil cleaning, thermal cracking (which breaks down tars at high enough temperatures, often above 1,000°C), and catalytic cracking [83, 84]. The generation of carcinogenic polyaromatic hydrocarbon may be facilitated by thermal cracking, which consumes a lot of energy and cannot crack refractory tars. The most advanced syngas cleaning method is thought to include catalytic cracking of tar. At low temperatures (less than 700°C) and with little energy, the right catalyst can be employed to provide excellent tar removal efficiency (more than 90%) during catalytic cracking.

6.4.4.2 A CATALYST FOR THE SYNTHESIS OF BIODIESEL

The synthesis of biodiesel typically entails the esterification and trans-esterification of vegetable oil or animal fat using homogeneous and heterogeneous acid catalysts. Biochar has been demonstrated to be a useful precursor for the creation of heterogeneous acid catalysts (also known as solid acid catalysts), provided the material is treated properly [85].

6.4.4.3 ADDITIONAL AGRICULTURAL AND ENVIRONMENTAL BENEFITS

Several agricultural uses of biochar, including pesticide remediation, microbial growth support, and disease management, have been investigated [86–88]. To maintain soil health, nutrient cycling, fertility, disease suppression, and crop yield, microbial communities are essential to the functioning of the plant-soil ecosystem [89]. Some environmental applications of nano-biochar, including carbon sequestration, wastewater treatment, mitigating climate change, energy production, and organic or inorganic pollutant removal, are shown in Figure 6.4. These carbon nanoparticles may offer a long-term remedy for soil contamination brought on by pesticides and may be a possibility for pesticide remediation due to their enhanced affinity for pesticides [90–106]. However, in the future, more studies and research are required to understand biochar nanoparticles and their potential to adsorb various pesticides and herbicides.

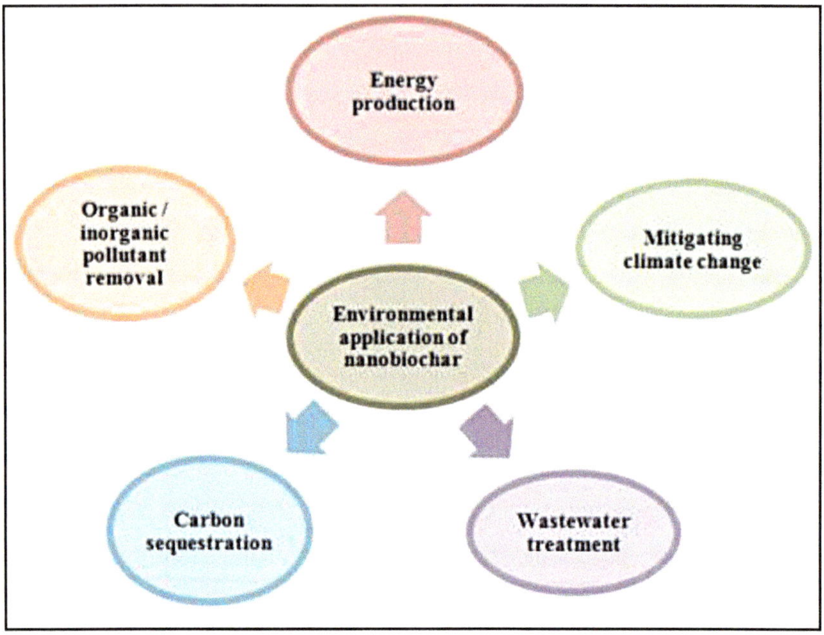

FIGURE 6.4 Environmental application of nano biochar.

6.5 CONCLUSION

Recent developments in biochar production, use, characteristics, and approaches are addressed. Biochar has been effectively used in a variety of applications, including soil amendment, as was previously mentioned. Potential uses for biochar include its usage as a catalyst precursor, sorbent for removing contaminants from soil and water, a sorbent for enhancing soil quality and carbon sequestration. Most of the research has shown how nanobiochar or biochar is used in agricultural and environmental problems. Biochar's intrinsic propensity to hold heavy metals can also make them more available to the soil. The ecotoxicological effects of biochar on soil organisms can also degrade. However, biochar qualities need to be improved and tailored for the right applications to be efficiently used for a variety of reasons. Overall, the utilization of char as a sustainable long-term product appears to have a very bright future.

KEYWORDS

- **agriculture application**
- **biochar**
- **cation exchange capacity**
- **dynamic soil organic matter**
- **electrical conductivity**
- **nanobiochar**
- **pyrolysis**
- **sustainable**

REFERENCES

1. Luo, M., Lin, H., He, Y., & Zhang, Y. (2020). The influence of corncob-based biochar on remediation of arsenic and cadmium in yellow soil and cinnamon soil. *Science of The Total Environment, 717*, 137014. https://doi.org/10.1016/j.scitotenv.2020.137014.
2. Routa, J., Kellomäki, S., & Strandman, H. (2012). Effects of forest management on total biomass production and CO_2 emissions from use of energy biomass of Norway spruce and Scots pine. *Bioenergy Research, 5*, 733–747. https://doi.org/10.1007/s12155-012-9183-5.
3. Mortensen, P. M., Grunwaldt, J.-D., Jensen, P. A., Knudsen, K. G., & Jensen, A. D. (2011). A review of catalytic upgrading of bio-oil to engine fuels. *Applied Catalysis A: General, 407*, 1–19. https://doi.org/10.1016/j.apcata.2011.08.046.
4. Swain, P. K., Das, L. M., & Naik, S. N. (2011). Biomass to liquid: A prospective challenge to research and development in the 21st century. *Renewable and Sustainable Energy Reviews, 15*, 4917–4933. https://doi.org/10.1016/j.rser.2011.07.061.
5. Mishra, A., Gupta, B., Kumar, N., Singh, R., Varma, A., & Thakur, I. S. (2020). Synthesis of calcite-based bio-composite biochar for enhanced biosorption and detoxification of chromium Cr (VI) by *Zhihengliuella* sp. ISTPL4. *Bioresource Technology, 307*, 123262. https://doi.org/10.1016/j.biortech.2020.123262.
6. Gupta, B., Mishra, A., Singh, R., & Thakur, I. S. (2021). Fabrication of calcite-based biocomposites for catalytic removal of heavy metals from electroplating industrial effluent. *Environmental Technology & Innovation, 21*, 101278. https://doi.org/10.1016/j.eti.2020.101278.
7. Hammer, E. C., Forstreuter, M., Rillig, M. C., & Kohler, J. (2015). Biochar increases arbuscular mycorrhizal plant growth enhancement and ameliorates salinity stress. *Applied Soil Ecology, 96*, 114–121. https://doi.org/10.1016/j.apsoil.2015.07.014.

8. Zhang, R., Zhang, Y., Song, L., Song, X., Hänninen, H., & Wu, J. (2017). Biochar enhances the nut quality of *Torreya grandis* and soil fertility under simulated nitrogen deposition. *Forest Ecology and Management, 391*, 321–329. https://doi.org/10.1016/j.foreco.2017.02.036.
9. Qiu, B., Tao, X., Wang, H., Li, W., Ding, X., & Chu, H. (2021). Biochar as a low-cost adsorbent for aqueous heavy metal removal: A review. *Journal of Analytical and Applied Pyrolysis, 155*, 105081. https://doi.org/10.1016/j.jaap.2021.105081.
10. He, M., Xiong, X., Wang, L., Hou, D., Bolan, N. S., Ok, Y. S., Rinklebe, J., & Tsang, D. C. W. (2021). A critical review of performance indicators for evaluating soil biota and soil health of biochar-amended soils. *Journal of Hazardous Materials, 414*, 125378. https://doi.org/10.1016/j.jhazmat.2021.125378.
11. Zech, W., Haumaier, L., & Hempfling, R. (2015). Ecological aspects of soil organic matter in tropical land use. In *Humic Substances in Soil and Crop Sciences: Selected Readings* (pp. 187–202). Soil Science Society of America.
12. Glaser, B., Haumaier, L., Guggenberger, G., & Zech, W. (2001). The "Terra Preta" phenomenon: A model for sustainable agriculture in the humid tropics. *Naturwissenschaften, 88*, 37–41. https://doi.org/10.1007/s001140000193.
13. Lehmann, J. (2007). A handful of carbon. *Nature, 447*, 143–144. https://doi.org/10.1038/447143a.
14. Alimohammadi, M., Xu, Y., Wang, D., Biris, A. S., & Khodakovskaya, M. V. (2011). Physiological responses induced in tomato plants by a two-component nanostructural system composed of carbon nanotubes conjugated with quantum dots and its in vivo multimodal detection. *Nanotechnology, 22*, 295101. https://doi.org/10.1088/0957-4484/22/29/295101.
15. Tripathi, S., Sonkar, S. K., & Sarkar, S. (2011). Growth stimulation of gram (*Cicer arietinum*) plant by water-soluble carbon nanotubes. *Nanoscale, 3*, 1176. https://doi.org/10.1039/c0nr00722f.
16. Ashfaq, M., Verma, N., & Khan, S. (2017). Carbon nanofibers as a micronutrient carrier in plants: Efficient translocation and controlled release of Cu nanoparticles. *Environmental Science: Nano, 4*, 138–148. https://doi.org/10.1039/C6EN00385K.
17. Qian, K., Kumar, A., Patil, K., Bellmer, D., Wang, D., Yuan, W., & Huhnke, R. (2013). Effects of biomass feedstocks and gasification conditions on the physiochemical properties of char. *Energies, 6*, 3972–3986. https://doi.org/10.3390/en6083972.
18. Kumar, A., Jones, D., & Hanna, M. (2009). Thermochemical biomass gasification: A review of the current status of the technology. *Energies, 2*, 556–581. https://doi.org/10.3390/en20300556.
19. Yaashikaa, P. R., Kumar, P. S., Varjani, S., & Saravanan, A. (2020). A critical review of the biochar production techniques, characterization, stability, and applications for circular bioeconomy. *Biotechnology Reports, 28*, e00570. https://doi.org/10.1016/j.btre.2020.e00570.
20. Lehmann, J., & Joseph, S. (Eds.). (2024). *Biochar for environmental management: science, technology and implementation*. Taylor & Francis, pp. 1–902.
21. Rousset, P., Macedo, L., Commandré, J.-M., & Moreira, A. (2012). Biomass torrefaction under different oxygen concentrations and its effect on the composition of the solid by-product. *Journal of Analytical and Applied Pyrolysis, 96*, 86–91. https://doi.org/10.1016/j.jaap.2012.03.009.

22. Bergman, P. C., Boersma, A. R., Zwart, R. W. R., & Kiel, J. H. A. (2005). Torrefaction for biomass co-firing in existing coal-fired power stations. *Energy Research Centre of the Netherlands, ECN-C-05-013, 31.*
23. Hoekman, S. K., Broch, A., Robbins, C., Zielinska, B., & Felix, L. (2013). Hydrothermal carbonization (HTC) of selected woody and herbaceous biomass feedstocks. *Biomass Conversion and Biorefinery, 3*, 113–126. https://doi.org/10.1007/s13399-012-0066-y.
24. Benavente, V., Calabuig, E., & Fullana, A. (2015). Upgrading of moist agro-industrial wastes by hydrothermal carbonization. *Journal of Analytical and Applied Pyrolysis, 113*, 89–98. https://doi.org/10.1016/j.jaap.2014.11.004.
25. Benedetti, V., Patuzzi, F., & Baratieri, M. (2017). Gasification char is a potential substitute for activated carbon in adsorption applications. *Energy Procedia, 105*, 712–717. https://doi.org/10.1016/j.egypro.2017.03.380.
26. Fahmi, A. H., Samsuri, A. W., Jol, H., & Singh, D. (2018). Physical modification of biochar to expose the inner pores and their functional groups to enhance lead adsorption. *RSC Advances, 8*, 38270–38280. https://doi.org/10.1039/C8RA06867D.
27. Li, G., Khan, S., Ibrahim, M., Sun, T.-R., Tang, J.-F., Cotner, J. B., & Xu, Y.-Y. (2018). Biochars induced modification of dissolved organic matter (DOM) in soil and its impact on mobility and bioaccumulation of arsenic and cadmium. *Journal of Hazardous Materials, 348*, 100–108. https://doi.org/10.1016/j.jhazmat.2018.01.031.
28. Santos, M. P., Brito, M. J., Junior, E. C., Bonomo, R. C., & Veloso, C. M. (2019). Pepsin immobilization on biochar by adsorption and covalent binding, and its application for hydrolysis of bovine casein. *Journal of Chemical Technology & Biotechnology, 94*, 1982–1990. https://doi.org/10.1002/jctb.5981.
29. Li, J., Cao, L., Yuan, Y., Wang, R., Wen, Y., & Man, J. (2018). Comparative study for microcystin-LR sorption onto biochars produced from various plant- and animal-wastes at different pyrolysis temperatures: Influencing mechanisms of biochar properties. *Bioresource Technology, 247*, 794–803. https://doi.org/10.1016/j.biortech.2017.09.120.
30. Yue, Y., Lin, Q., Xu, Y., Li, G., & Zhao, X. (2017). Slow pyrolysis is a measure for rapidly treating cow manure and its biochar characteristics. *Journal of Analytical and Applied Pyrolysis, 124*, 355–361. https://doi.org/10.1016/j.jaap.2017.01.008.
31. Angın, D. (2013). Effect of pyrolysis temperature and heating rate on biochar obtained from pyrolysis of safflower seed press cake. *Bioresource Technology, 128*, 593–597. https://doi.org/10.1016/j.biortech.2012.10.150.
32. Liang, B., Lehmann, J., Solomon, D., Kinyangi, J., Grossman, J., O'Neill, B., Skjemstad, J. O., Thies, J., Luizão, F. J., Petersen, J., & Neves, E. G. (2006). Black carbon increases cation exchange capacity in soils. *Soil Science Society of America Journal, 70*, 1719–1730. https://doi.org/10.2136/sssaj2005.0383.
33. Liu, C., Wang, H., Tang, X., Guan, Z., Reid, B. J., Rajapaksha, A. U., Ok, Y. S., & Sun, H. (2016). Biochar increased water holding capacity but accelerated organic carbon leaching from a sloping farmland soil in China. *Environmental Science and Pollution Research, 23*, 995–1006. https://doi.org/10.1007/s11356-015-4885-9.
34. Balaganesh, B. P. M. C. V. (2014). Impact of biochar on soil health. *International Journal of Advanced Research (Indore), 2*, 933–950.

35. Berihun, T., Tadele, M., & Kebede, F. (2017). The application of biochar on soil acidity and other physico-chemical properties of soils in southern Ethiopia. *Journal of Plant Nutrition and Soil Science, 180*, 381–388. https://doi.org/10.1002/jpln.201600343.
36. Githinji, L. (2014). Effect of biochar application rate on soil physical and hydraulic properties of a sandy loam. *Archives of Agronomy and Soil Science, 60*, 457–470. https://doi.org/10.1080/03650340.2013.821698.
37. Mukherjee, A., Lal, R., & Zimmerman, A. R. (2014). Effects of biochar and other amendments on the physical properties and greenhouse gas emissions of artificially degraded soil. *Science of The Total Environment, 487*, 26–36. https://doi.org/10.1016/j.scitotenv.2014.03.141.
38. Cañasveras, J. C., Barrón, V., del Campillo, M. C., Torrent, J., & Gómez, J. A. (2010). Estimation of aggregate stability indices in Mediterranean soils by diffuse reflectance spectroscopy. *Geoderma, 158*, 78–84. https://doi.org/10.1016/j.geoderma.2009.09.004.
39. Asai, H., Samson, B. K., Stephan, H. M., Songyikhangsuthor, K., Homma, K., Kiyono, Y., Inoue, Y., Shiraiwa, T., & Horie, T. (2009). Biochar amendment techniques for upland rice production in northern Laos. *Field Crops Research, 111*, 81–84. https://doi.org/10.1016/j.fcr.2008.10.008.
40. Tan, Z., Lin, C. S. K., Ji, X., & Rainey, T. J. (2017). Returning biochar to fields: A review. *Applied Soil Ecology, 116*, 1–11. https://doi.org/10.1016/j.apsoil.2017.03.017.
41. Munda, S., Bhaduri, D., Mohanty, S., Chatterjee, D., Tripathi, R., Shahid, M., Kumar, U., Bhattacharyya, P., Kumar, A., Adak, T., Jangde, H. K., & Nayak, A. K. (2018). Dynamics of soil organic carbon mineralization and C fractions in paddy soil on application of rice husk biochar. *Biomass and Bioenergy, 115*, 1–9. https://doi.org/10.1016/j.biombioe.2018.04.002.
42. Helal, M. E. H., & Mohamed, I. D. (2019). Characterization of agricultural residues-based nano biochar and its efficiency in adsorption/desorption of nutrients. *International Journal of Environment, 8*(2), 130–141.
43. Wang, S., Dai, G., Yang, H., & Luo, Z. (2017). Lignocellulosic biomass pyrolysis mechanism: A state-of-the-art review. *Progress in Energy and Combustion Science, 62*, 33–86. https://doi.org/10.1016/j.pecs.2017.05.004.
44. Alburquerque, J. A., Salazar, P., Barrón, V., Torrent, J., del Campillo, M. C., Gallardo, A., & Villar, R. (2013). Enhanced wheat yield by biochar addition under different mineral fertilization levels. *Agronomy for Sustainable Development, 33*(3), 475–484. https://doi.org/10.1007/s13593-012-0128-3.
45. Liu, X., Zhang, A., Ji, C., Joseph, S., Bian, R., Li, L., Pan, G., & Paz-Ferreiro, J. (2013). Biochar's effect on crop productivity and the dependence on experimental conditions—A meta-analysis of literature data. *Plant and Soil, 373*(1), 583–594. https://doi.org/10.1007/s11104-013-1806-x.
46. Karami, N., Clemente, R., Moreno-Jiménez, E., Lepp, N. W., & Beesley, L. (2011). Efficiency of green waste compost and biochar soil amendments for reducing lead and copper mobility and uptake to ryegrass. *Journal of Hazardous Materials, 191*(1–3), 41–48. https://doi.org/10.1016/j.jhazmat.2011.04.025.
47. Xiu, S., Shahbazi, A., & Li, R. (2017). Characterization, modification, and application of biochar for energy storage and catalysis: A review. *Trends in Renewable Energy, 3*(1), 86–101. https://doi.org/10.17737/tre.2017.3.1.0033.

48. Ramadan, M. M., Asran-Amal, & Abd-Elsalam, K. A. (2020). Micro/nano biochar for sustainable plant health: Present status and prospects. In *Carbon Nanomaterials for Agri-Food and Environmental Applications* (pp. 323–357). Elsevier.
49. Bah, A., Husni, M. H. A., Teh, C. B. S., Rafii, M. Y., Syed Omar, S. R., & Ahmed, O. H. (2014). Reducing runoff loss of applied nutrients in oil palm cultivation using controlled-release fertilizers. *Advances in Agriculture, 2014*, 1–9. https://doi.org/10.1155/2014/285387.
50. Withers, P. J. A., Clay, S. D., & Breeze, V. G. (2001). Phosphorus transfer in runoff following application of fertilizer, manure, and sewage sludge. *Journal of Environmental Quality, 30*(1), 180–188. https://doi.org/10.2134/jeq2001.301180x.
51. Sohi, S. P., Krull, E., Lopez-Capel, E., & Bol, R. (2010). A review of biochar and its use and function in soil. In *Advances in Agronomy* (Vol. 105, pp. 47–82). Elsevier.
52. Debicka, M., Kocowicz, A., Weber, J., & Jamroz, E. (2016). Organic matter effects on phosphorus sorption in sandy soils. *Archives of Agronomy and Soil Science, 62*(6), 840–855. https://doi.org/10.1080/03650340.2015.1083981.
53. Rawat, J., Saxena, J., & Sanwal, P. (2019). Biochar: A Sustainable Approach for Improving Plant Growth and Soil Properties [Internet]. Biochar - An Imperative Amendment for Soil and the Environment. IntechOpen; Available from: http://dx.doi.org/10.5772/intechopen.82151.
54. Saxena, M., Maity, S., & Sarkar, S. (2014). Carbon nanoparticles in 'biochar' boost wheat (*Triticum aestivum*) plant growth. *RSC Advances, 4*(75), 39948–39954. https://doi.org/10.1039/C4RA06535B.
55. Steiner, C., Glaser, B., Geraldes Teixeira, W., Lehmann, J., Blum, W. E. H., & Zech, W. (2008). Nitrogen retention and plant uptake on a highly weathered central Amazonian Ferralsol amended with compost and charcoal. *Journal of Plant Nutrition and Soil Science, 171*(6), 893–899. https://doi.org/10.1002/jpln.200625199.
56. Warnock, D. D., Lehmann, J., Kuyper, T. W., & Rillig, M. C. (2007). Mycorrhizal responses to biochar in soil – Concepts and mechanisms. *Plant and Soil, 300*(1–2), 9–20. https://doi.org/10.1007/s11104-007-9391-5.
57. Yu, X.-Y., Mu, C.-L., Gu, C., Liu, C., & Liu, X.-J. (2011). Impact of woodchip biochar amendment on the sorption and dissipation of pesticide acetamiprid in agricultural soils. *Chemosphere, 85*(8), 1284–1289. https://doi.org/10.1016/j.chemosphere.2011.07.031.
58. Gwenzi, W., Nyambishi, T. J., Chaukura, N., & Mapope, N. (2018). Synthesis and nutrient release patterns of a biochar-based N–P–K slow-release fertilizer. *International Journal of Environmental Science and Technology, 15*(2), 405–414. https://doi.org/10.1007/s13762-017-1399-7.
59. Hossain, M. K., Strezov, V., Chan, K. Y., Ziolkowski, A., & Nelson, P. F. (2011). Influence of pyrolysis temperature on production and nutrient properties of wastewater sludge biochar. *Journal of Environmental Management, 92*(1), 223–228. https://doi.org/10.1016/j.jenvman.2010.09.008.
60. Bopp, C., Christl, I., Schulin, R., & Evangelou, M. W. H. (2016). Biochar as a possible long-term soil amendment for phytostabilisation of TE-contaminated soils. *Environmental Science and Pollution Research, 23*(17), 17449–17458. https://doi.org/10.1007/s11356-016-6935-3.

61. Lateef, A., Nazir, R., Jamil, N., Alam, S., Shah, R., Khan, M. N., Saleem, M., & Rehman, S. (2019). Synthesis and characterization of environmentally friendly corncob biochar-based nano-composite – A potential slow-release nano-fertilizer for sustainable agriculture. *Environmental Nanotechnology, Monitoring & Management, 11*, 100212. https://doi.org/10.1016/j.enmm.2019.100212.
62. Chen, L., Chen, Q., Rao, P., Yan, L., Shakib, A., & Shen, G. (2018). Formulating and optimizing a novel biochar-based fertilizer for simultaneous slow-release of nitrogen and immobilization of cadmium. *Sustainability, 10*(8), 2740. https://doi.org/10.3390/su10082740.
63. Dominguez, E. L., Uttran, A., Loh, S. K., Manero, M.-H., Upperton, R., Idris Tanimu, M., & Thomas Bachmann, R. (2020). Characterization of industrially produced oil palm kernel shell biochar and its potential as slow-release nitrogen-phosphate fertilizer and carbon sink. *Materials Today: Proceedings, 31*, 221–227. https://doi.org/10.1016/j.matpr.2020.05.143.
64. Khodakovskaya, M. V., de Silva, K., Biris, A. S., Dervishi, E., & Villagarcia, H. (2012). Carbon nanotubes induce growth enhancement of tobacco cells. *ACS Nano, 6*(3), 2128–2135. https://doi.org/10.1021/nn204643g.
65. Lin, S., Reppert, J., Hu, Q., Hudson, J. S., Reid, M. L., Ratnikova, T. A., Rao, A. M., Luo, H., & Ke, P. C. (2009). Uptake, translocation, and transmission of carbon nanomaterials in rice plants. *Small, 5*(10), 1128–1132. https://doi.org/10.1002/smll.200801556.
66. Sonkar, S. K., Roy, M., Babar, D. G., & Sarkar, S. (2012). Water soluble carbon nano-onions from wood wool as growth promoters for gram plants. *Nanoscale, 4*(24), 7670–7675. https://doi.org/10.1039/c2nr32408c.
67. Khodakovskaya, M. V., de Silva, K., Nedosekin, D. A., Dervishi, E., Biris, A. S., Shashkov, E. V., Galanzha, E. I., & Zharov, V. P. (2011). Complex genetic, photothermal, and photoacoustic analysis of nanoparticle-plant interactions. *Proceedings of the National Academy of Sciences, 108*(3), 1028–1033. https://doi.org/10.1073/pnas.1008856108.
68. Wang, X., Han, H., Liu, X., Gu, X., Chen, K., & Lu, D. (2012). Multi-walled carbon nanotubes can enhance the root elongation of wheat (*Triticum aestivum*) plants. *Journal of Nanoparticle Research, 14*, 841. https://doi.org/10.1007/s11051-012-0841-5.
69. Wu, L., Wei, C., Zhang, S., Wang, Y., Kuzyakov, Y., & Ding, X. (2019). MgO-modified biochar increases phosphate retention and rice yields in saline-alkaline soil. *Journal of Cleaner Production, 235*, 901–909. https://doi.org/10.1016/j.jclepro.2019.07.043.
70. Zhao, Q., Ma, C., White, J. C., Dhankher, O. P., Zhang, X., Zhang, S., & Xing, B. (2017). Quantitative evaluation of multi-wall carbon nanotube uptake by terrestrial plants. *Carbon, 114*, 661–670. https://doi.org/10.1016/j.carbon.2016.12.036.
71. Ahmad, M., Rajapaksha, A. U., Lim, J. E., Zhang, M., Bolan, N., Mohan, D., Vithanage, M., Lee, S. S., & Ok, Y. S. (2014). Biochar as a sorbent for contaminant management in soil and water: A review. *Chemosphere, 99*, 19–33. https://doi.org/10.1016/j.chemosphere.2013.10.071.
72. Chen, B., Zhou, D., & Zhu, L. (2008). Transitional adsorption and partition of nonpolar and polar aromatic contaminants by biochars of pine needles with different pyrolytic temperatures. *Environmental Science & Technology, 42*(14), 5137–5143. https://doi.org/10.1021/es8002684.

73. Zhang, M., Gao, B., Varnoosfaderani, S., Hebard, A., Yao, Y., & Inyang, M. (2013). Preparation and characterization of a novel magnetic biochar for arsenic removal. *Bioresource Technology, 130,* 457–462. https://doi.org/10.1016/j.biortech.2012.11.132.
74. Naghdi, M., Taheran, M., Pulicharla, R., Rouissi, T., Brar, S. K., Verma, M., & Surampalli, R. Y. (2019). Pine-wood derived nano biochar for removal of carbamazepine from aqueous media: Adsorption behavior and influential parameters. *Arabian Journal of Chemistry, 12*(8), 5292–5301. https://doi.org/10.1016/j.arabjc.2016.12.025.
75. Yao, Y., Gao, B., Fang, J., Zhang, M., Chen, H., Zhou, Y., Creamer, A. E., Sun, Y., & Yang, L. (2014). Characterization and environmental applications of clay–biochar composites. *Chemical Engineering Journal, 242,* 136–143. https://doi.org/10.1016/j.cej.2013.12.062.
76. Zhang, M., Gao, B., Yao, Y., Xue, Y., & Inyang, M. (2012). Synthesis, characterization, and environmental implications of graphene-coated biochar. *Science of the Total Environment, 435–436,* 567–572. https://doi.org/10.1016/j.scitotenv.2012.07.038.
77. Wang, M. C., Sheng, G. D., & Qiu, Y. P. (2015). A novel manganese-oxide/biochar composite for efficient removal of lead (II) from aqueous solutions. *International Journal of Environmental Science and Technology, 12,* 1719–1726. https://doi.org/10.1007/s13762-014-0538-7.
78. Feng, Y., Lu, H., Liu, Y., Xue, L., Dionysiou, D. D., Yang, L., & Xing, B. (2017). Nano-cerium oxide functionalized biochar for phosphate retention: Preparation, optimization, and rice paddy application. *Chemosphere, 185,* 816–825. https://doi.org/10.1016/j.chemosphere.2017.07.107.
79. He, R., Peng, Z., Lyu, H., Huang, H., Nan, Q., & Tang, J. (2018). Synthesis and characterization of an iron-impregnated biochar for aqueous arsenic removal. *Science of the Total Environment, 612,* 1177–1186. https://doi.org/10.1016/j.scitotenv.2017.09.016.
80. Lian, F., Yu, W., Zhou, Q., Gu, S., Wang, Z., & Xing, B. (2020). Size matters: Nano-biochar triggers decomposition and transformation inhibition of antibiotic resistance genes in aqueous environments. *Environmental Science & Technology, 54*(15), 8821–8829. https://doi.org/10.1021/acs.est.0c02227.
81. Li, R., Zhang, Y., Deng, H., Zhang, Z., Wang, J. J., Shaheen, S. M., Xiao, R., Rinklebe, J., Xi, B., He, X., & Du, J. (2020). Removing tetracycline and Hg (II) with ball-milled magnetic nano biochar and its potential on polluted irrigation water reclamation. *Journal of Hazardous Materials, 384,* 121095. https://doi.org/10.1016/j.jhazmat.2019.121095.
82. Li, C., Zhang, L., Gao, Y., & Li, A. (2018). Facile synthesis of nano ZnO/ZnS modified biochar by directly pyrolyzing zinc-contaminated corn stover for Pb (II), Cu (II), and Cr (VI) removals. *Waste Management, 79,* 625–637. https://doi.org/10.1016/j.wasman.2018.08.035.
83. Mudinoor, A. R., Bellmer, D., Marin, L., Kumar, A., & Huhnke, R. (2011). Conversion of toluene (model tar) using selected steam reforming catalysts. *Transactions of the ASABE, 54,* 1819–1827. https://doi.org/10.13031/2013.39821.

84. Anis, S., & Zainal, Z. A. (2011). Tar reduction in biomass producer gas via mechanical, catalytic and thermal methods: A review. *Renewable and Sustainable Energy Reviews, 15,* 2355–2377. https://doi.org/10.1016/j.rser.2011.02.018.
85. Li, M., Zheng, Y., Chen, Y., & Zhu, X. (2014). Biodiesel production from waste cooking oil using a heterogeneous catalyst from pyrolyzed rice husk. *Bioresource Technology, 154,* 345–348. https://doi.org/10.1016/j.biortech.2013.12.070.
86. Saxena, J., Rawat, J., & Sanwal, P. (2016). Enhancement of growth and yield of Glycine max plants with inoculation of phosphate solubilizing fungus *Aspergillus niger* K7 and biochar amendment in soil. *Communications in Soil Science and Plant Analysis, 47,* 2334–2347. https://doi.org/10.1080/00103624.2016.1243708.
87. Thies, J. E. (2015). Biochar effects on the abundance, activity and diversity of the soil biota. In *Biochar for environmental management* (pp. 327–389). Routledge.
88. Cernansky, R. (2015). Agriculture: State-of-the-art soil. *Nature, 517,* 258–260. https://doi.org/10.1038/517258a.
89. Rajput, V. D., Minkina, T., Kumari, A., Harish, Singh, V. K., Verma, K. K., Mandzhieva, S., Sushkova, S., Srivastava, S., & Keswani, C. (2021). Coping with the challenges of abiotic stress in plants: New dimensions in the field application of nanoparticles. *Plants, 10*(6), 1221. https://doi.org/10.3390/plants10061221.
90. Rajput, V., Minkina, T., Sushkova, S., Behal, A., Maksimov, A., Blicharska, E., Ghazaryan, K., Movsesyan, H., & Barsova, N. (2019). ZnO and CuO nanoparticles: A threat to soil organisms, plants, and human health. *Environmental Geochemistry and Health, 42*(1), 147–158. https://doi.org/10.1007/s10653-019-00317-3.
91. Rajput, V. D., Minkina, T., Sushkova, S., Tsitsuashvili, V., Mandzhieva, S., Gorovtsov, A., Nevidomskyaya, D., & Gromakova, N. (2017). Effect of nanoparticles on crops and soil microbial communities. *Journal of Soils and Sediments, 18*(6), 2179–2187. https://doi.org/10.1007/s11368-017-1793-2.
92. Rajput, V. D., Minkina, T., Fedorenko, A., Sushkova, S., Mandzhieva, S., Lysenko, V., Duplii, N., Fedorenko, G., Dvadnenko, K., & Ghazaryan, K. (2018). Toxicity of copper oxide nanoparticles on spring barley (*Hordeum sativum* distichum). *Science of The Total Environment, 645,* 1103–1113. https://doi.org/10.1016/j.scitotenv.2018.07.211.
93. Rajput, V. D., Kumari, A., Minkina, T., Barakhov, A., Singh, S., Mandzhieva, S. S., Sushkova, S., Ranjan, A., Rajput, P., & Garg, M. C. (2022). A practical evaluation of the integrated role of biochar and nanomaterials in soil remediation processes. *Environmental Geochemistry and Health, 45, 9435–9449.* https://doi.org/10.1007/s10653-022-01375-w.
94. Rajput, V. D., Faizan, M., Upadhyay, S. K., Kumari, A., Ranjan, A., Sushkova, S., Chauhan, P. K., Mahmud, A. A., Burachevskaya, M., Chaplygin, V., & Deryabkina, I. (2022). Influence of nanoparticles on the plant rhizosphere microbiome. In *Sustainable Plant Nutrition in a Changing World* (pp. 83–102). https://doi.org/10.1007/978-3-030-97389-6_4.
95. Rajput, V. D., Singh, A., Minkina, T., Rawat, S., Mandzhieva, S., Sushkova, S., Shuvaeva, V., Nazarenko, O., Rajput, P., Komariah, Verma, K. K., Singh, A. K., Rao, M., & Upadhyay, S. K. (2021). Nano-enabled products: Challenges and opportunities for sustainable agriculture. *Plants, 10*(12), 2727. https://doi.org/10.3390/plants10122727.

96. Singh, A., Singh, A. K., Rawat, S., Pal, N., Rajput, V. D., Minkina, T., Sharma, R., Singh, N. P., & Tripathi, J. N. (2022). Satellite-based quantification of methane emissions from wetlands and rice paddies ecosystems in North and Northeast India. *Hydrobiology, 1*(3), 317–330. https://doi.org/10.3390/hydrobiology1030023.
97. Singh, A., Sengar, R. S., Rajput, V. D., Minkina, T., & Singh, R. K. (2022). Zinc oxide nanoparticles improve salt tolerance in rice seedlings by improving physiological and biochemical indices. *Agriculture, 12*(7), 1014. https://doi.org/10.3390/agriculture12071014.
98. Singh, A., Sengar, R. S., Sharma, R., & Rajput, P. (2021). Nano-priming technology for sustainable agriculture. *Biogeosystem Technique, 8*(2), 79–92. https://doi.org/10.13187/bgt.2021.2.79.
99. Singh, A., Mehrotra, R., Rajput, V. D., Dmitriev, P., Singh, A. K., Kumar, P., Tomar, R. S., Singh, O., & Singh, A. K. (2022). Geoinformatics, artificial intelligence, sensor technology, big data. In *Sustainable Agriculture Systems and Technologies* (pp. 295–313). https://doi.org/10.1002/9781119808565.ch14.
100. Singh, A., Rajput, V. D., Rawat, S., Sharma, R., Singh, A. K., Kumar, P., Singh, A. K., Minkina, T., Singh, R. P., & Singh, S. (2022). Geoinformatics and nanotechnological approaches for coping up abiotic and biotic stress in crop plants. In *Sustainable Agriculture Systems and Technologies* (pp. 337–359). https://doi.org/10.1002/9781119808565.ch17.
101. Singh, A., Rajput, V. D., Rawat, S., Sharma, R., Kumar Singh, A., Kumar Singh, A., Tomar, R. S. (2021). Emerging tools for sustainable agriculture and food security: An introduction. In Deepika (Ed.), *Emerging Tools for Sustainable Agriculture and Food Security, 1st ed.;* New Delhi: Deepika, *1*(1), 3–16.
102. Rajput, P., Rajput, V. D., Singh, A., Sharma, R., Rawat, S., Tomar, R. S., Singh, A., & Minkina, T. (2021). Impact of emerging tools for solving 'hidden hunger' in humans. *Emerging Tools for Sustainable Agriculture and Food Security, 1*(1), 17–30.
103. Rajput, V. D., Minkina, T., Kumari, A., Shende, S. S., Ranjan, A., Faizan, M., Barakvov, A., Gromovik, A., Gorbunova, N., Rajput, P., Singh, A., Khabirov, I., Nazarenko, O., Sushkova, S., & Kizilkaya, R. (2022). A review on nanobioremediation approaches for restoration of contaminated soil. *Eurasian Journal of Soil Science (EJSS), 11*(1), 43–60. https://doi.org/10.18393/ejss.990605.
104. Rajput, V. D., Minkina, T., Feizi, M., Kumari, A., Khan, M., Mandzhieva, S., Sushkova, S., El-Ramady, H., Verma, K. K., Singh, A., van Hullebusch, E. D., Singh, R. K., Jatav, H. S., & Choudhary, R. (2021). Effects of silicon and silicon-based nanoparticles on rhizosphere microbiome, plant stress and growth. *Biology, 10*(8), 791. https://doi.org/10.3390/biology10080791.
105. Singh, A., et al. (2020). Monitoring soil salinity and recent advances in the mechanism of salinity tolerance in plants. *Biogeosystem Technique, 7*(2), 66–87. https://doi.org/10.13187/bgt.2020.2.66.
106. Singh, A., Sengar, R. S., Shahi, U. P., Rajput, V. D., Minkina, T., & Ghazaryan, K. A. (2022). Prominent effects of zinc oxide nanoparticles on roots of rice (*Oryza sativa* L.) grown under salinity stress. *Stresses, 3,* 33–46. MDPI AG. https://doi.org/10.3390/stresses3010004.

CHAPTER 7

Insights into Nanoparticles and Nano-Biochar in Soil-Microbe Interaction

RITIKA CHAUHAN, GAURAV YADAV, AJIT VARMA, D. K. CHOUDHARY, and ARTI GOEL

Amity Institute of Microbial Technology, Amity University, Noida, Uttar Pradesh, India

ABSTRACT

Advanced nanostructured materials, including nano-biochar and nanoparticles, have provided long-term fixes for a number of contemporary challenges. Nano-biochar and biochar nanocomposites may be produced as a result of biochar technology and nanotechnology, revolutionizing research in the relevant domains. Nano-biochar is a kind of biochar with improved surface, chemical, and physical properties. It has many benefits, including enhancing soil quality and plant development, managing plant diseases, and bioremediating pesticides and pollutants. The organic character of biochar produced at low pyrolysis temperatures is more varied and includes aliphatic and cellulose-type compounds. These serve as advantageous substrates for the mineralization of bacteria and fungi, which are essential for the processes of nutrient turnover and aggregate formation. The addition of biochar decreased N_2O and CO_2 emissions. The diversity and microbial community were altered by the addition of biochar. In this chapter, we have discussed the prospects of nano and nano-biochar by altering soil microbial flora and other different beneficial parameters for sustainable agriculture.

Nano-Biochar and Nanoparticles for Soil Health Improvement.
Vishnu D. Rajput and Tatiana M. Minkina (Eds.)

7.1 INTRODUCTION

Different countries have employed a variety of distinctive ways for reduction, repurposing, composting, combustion, and disposal into landfills to manage compacted waste. According to the World Bank and the Agency for International Development (USAID), a municipal council in developing countries spends 20% to 50% of its financial budget on reducing solid trash; these funds come from fewer than 50% of the inner-city population [1, 2]. According to World Bank survey statistics, around 1.3 billion tons of municipal solid trash are produced annually by cities throughout the world, and that number is projected to rise to 2.2 billion tons by 2025. Population growth, growing urbanization, and socioeconomic development in low- and core-earnings nations are the main causes of this [3]. Some agricultural practices have an impact on waste generation, which may seriously harm ecosystems, biodiversity, and natural resources as well as the health of people and animals [4]. The management of municipal solid waste today causes a number of ecological problems, such as enhanced energy and heat expenditure or the generation of a large amount of recyclable materials [5]. The rapid growth in both human and animal populations, industrialization, and migration from rural to urban areas due to economic growth are the main causes of the rapid increase of municipal trash [6]. A pyrogenic black carbon known as biochar is produced when biomass feedstocks, such as agricultural and forestry residues, are thermally converted in an inert environment. Due to its potential to reduce climate change and improve soil fertility and health, biochar technology has garnered a lot of interest [7]. Additionally, numerous studies have shown that biochar may be used in place of adsorbents to remove particular types of pollutants from aqueous solutions, such as heavy metals, vitamins, minerals, and pharmaceutical drugs [8]. Utilizing biochar might result in soils that begin to imitate the fertile terra preta, a variety of charcoal used to increase soil fertility [9]. But it has subsequently been proven that the unintended addition of inorganic (like ash and fish bones) and natural (such as biomass wastes, manure, excrements, urine, and biochar) components to extremely weathered and infertile Ferralsols is what actually caused terra preta to form [10]. It was first suggested to produce biochar for soil application in order to transfer waste biomass C from a fast to a slow C-cycling pond in soil [11]. An extensive study has been done in the past 10 years to understand additional possible consequences of biochar from their soil functions, in addition to the biochar's long-term C potential. For instance, biochar may reduce greenhouse gas emissions, boost the output of agricultural goods, and enhance soil characteristics and microbial, fungal, and mycorrhizal development [7, 12]. Despite having organic and natural origins, biochar is impervious to deterioration because of its extremely aromatic carbon structure. Due to its ability to boost cation exchange capacity [14], nutrient availability, and soil water-holding capacity, biochar continues to be one of the key plant-promoting components in agriculture. This chapter describes the safe application of micro/nano biochar for sustainable agriculture, including the bioremediation of various pesticides, antibacterial activities, controlled-release fertilizers, plant growth promoters, fragments, and silage agents. Because it is often a low-cost alternative to increase plant health and agricultural productivity, biochar has a very promising future.

7.2 DIFFERENCE BETWEEN BIOCHAR AND NANO-BIOCHAR

7.2.1 BIOCHAR

A solid substance called biochar is synthesized when biomass resources undergo thermochemical breakdown. Biochar is described as "a solid substance derived from the carbonization of biomass" by the International Biochar Initiative. The carbon-rich steady by-product of biomass carbonization, also known as "pyrochar," is created by heating wood, dung, or maybe leaves to temperatures between 300 and 1,000°C while maintaining a low oxygen content. The main evidence that charcoal was intentionally used by humans indicates that it was used as fuel in Southern Europe and the Middle East at least 5,500 years ago. In Britain, the Bronze Age began roughly 4,000 years ago, when charcoal was widely used as a metallurgical fuel. There is a lot of proof that Amazon Basin tribes utilized charcoal as a soil amendment together with dung, bone, and pottery shards to transform barren soils into ones that were lush and fruitful before the arrival of the Spanish conquistadors [15]. Due to the potential for a considerable reduction in atmospheric greenhouse gas concentrations, biochar has gained more attention in recent years. Peter identified biochar as a carbon-rich solid byproduct of pyrolysis, which could be the heat breakdown of biomass when oxygen is scarce. In the Amazon Delta Dark Earth, which has a greater organic C content and better fertility than the nearby organic (unaltered) soils, observations were made that gave birth to the notion of applying biochar to soil [16]. Biochar is one of the most common bio-sorbents. It is a C-rich adsorbent that has a high ability and efficacy to adsorb HMs from the environment owing to its high specific surface area, well-built porous construction, and high thermal stability. Biochar can be acquired by biological waste pyrolysis temperature ranges from 350°C to 800°C under oxygen-limited conditions.

Biomass includes not just living things, such as plants and animals, but also their waste products, sludge, and decomposing wood [17]. There are a few techniques that convert biomass components into syngas, bio-oil, and biochar, as well as synthesis gas that contains hydrogen, carbon monoxide, and a few other carbon dioxides. Due to the partial combustion of plant debris in pits, they form biochar, also known to us as terra preta. It was found that terra preta may maintain soil fertility for a very long time [18]. In general, biochar is made with long-term carbon storage in soil in mind and may also have positive effects on soil's physical properties, biological fertility, and chemical composition. Since it contains hydrogen (H), oxygen (O), nitrogen (N), sulfur (S), potassium (K), carbon (C), magnesium (Mg), and ash, biochar isn't made of carbon. The most crucial nourishment for plant development may be provided by these components. By increasing the amount of organic matter in the soil (such as organic carbon), biochar enhances the soil's physicochemical and biological qualities [19]. In particular, biochar has a higher surface area and porosity than other chemicals or biological changes, which enables us to adsorb nutrients, preserve water, or even provide habitats for beneficial soil microorganisms for the bioremediation of contaminated soil and drinking water and enhancement of soil health, crop yields, and controls phytopathogens (Figure 7.1) [20]. It has been shown that adding biochar to soil may significantly reduce greenhouse gas emissions from agricultural soils, such as an 80% reduction in N_2O emissions and complete elimination of methane generation [21].

7.3.2 NANO-BIOCHAR

It has been demonstrated that combining nanoparticles with plants for agrifood applications provides a wide range of advantages [22] In reality, due to their inherent ability to cross the cell membrane among various mammalian and plant cells and serve as vehicles for the delivery of drugs and genes, carbon nanotubes have attracted the attention of plant researchers as prospective molecular transporters [23]. Additionally, the production of biochar by thermochemical biomass processing can result

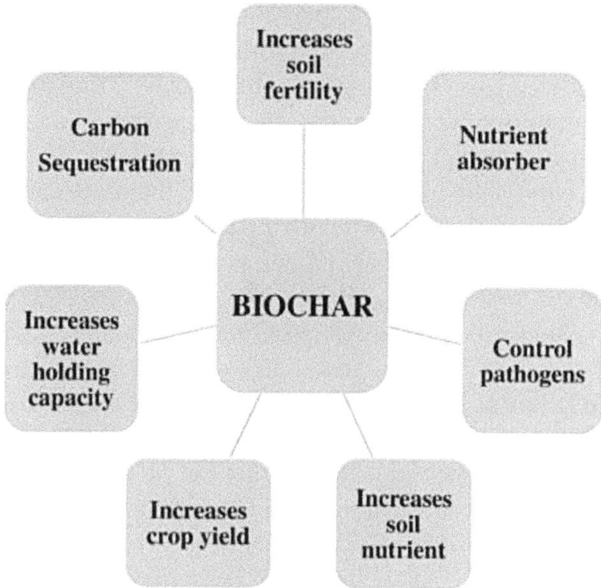

FIGURE 7.1 Biochar applications in plant and soil.

in the production of biofuels and syngas [24]. As a result, synthesizing biochar nanoparticles may achieve four integrated aims, namely the elimination of pollutants, the recycling of waste, the sequestration of carbon, and the generation of energy [25]. All these advantages point to the newest nanocomposite as a potential new solution and inventive adsorbent that may remove Cr (VI) from aqueous solutions inexpensively and effectively (with an optimal adsorption ability of 150.02 mg g_1). The progressive pyrolysis of dip-coated hickory or torrefied biomass in carbon nanotube (CNT) suspension systems is one way that might be used to generate the modified biochar nanocomposites [8]. Recent investigations have also produced a useful composite by impregnating biomass with functional nanoparticles before pyrolysis. For instance, by sluggishly pyrolyzing graphene-pretreated wheat straw biomass, graphene-enhanced the biochar composite. The combined findings showed that graphene was integrated into biochar's surface, generating a larger surface area, greater thermal stability, additional functional groups, and a higher efficiency at removing phenanthrene and mercury than untreated biochar [26]. Producing nanocomposites based on biochar would essentially combine the advantages of biochar with other nanomaterials, which would be quite important.

The functional groups, pore characteristics, surface-active locations, and catalytic devastation efficiency of the formed composites typically show dramatic improvements. They also separate quickly [27] in different ways. According to multiple recent studies, the process utilized to make nano biochar, such as controlling the pyrolyzing temperature or perhaps by exfoliation, controls the biochar particle size in the nanoscale range [28]. According to recent studies, the scale of passivated biochar was less than 10 nm. After pretreatment with $FeCl_3$, oil palm fruit bunches and biomass were carbonized to create magnetic nanobiochar composites. Using oleic acid, sulfuric acid was used to sulfonate magnetic nanobiochar, boosting its acidity ($MBCSO_3H$). Recent research discoveries open up new avenues for the manufacture of magnetic homogeneous acid catalysts from biomass waste [29]. A planetary ball mill was used to make nano-biochar from pine wood biochar. When biochar was kept at 80°C before milling, the particle size was reduced to 60 nm, and the surface area of nano-biochar was increased by 15 times when compared to raw biochar. Uses of the created nano-biochar is to remove the herbicide carbamazepine from wastewater through adsorption. Different requirements for employing biochar as the substrate for the development of nanocomposite materials are acknowledged, in contrast to diverse nanocomposite materials. First off, agricultural biomass and solid waste serve as plentiful and affordable feedstocks for the synthesis of biochar [30]. Second, producing biochar requires less energy, which is frequently created at very low temperatures (700°C) [31]. Micro/nano biochar can therefore be applied in a variety of ways, such as plant security, pesticide remediation, plant campaigns, etc. The extensive number of inspections attest to this [32, 33] that consistently produce large and rewarding benefits in plant systems.

7.4 TYPES OF NANO-BIOCHAR

Nanobiochar is divided into three different categories based on the nano-materials it contains: functional nanoparticles coated biochar, magnetic nanocomposites, and oxide/hydroxide biochar nanocomposites (e.g., graphene, graphene oxide, chitosan, carbon nanotubes, ZnS nanocrystals, etc.). Different synthesis processes are used to create these composites. Each synthesis technique involves the two treatment procedures, pre-treatment and post-treatment, of the biomass [27]:

1. **Nanocomposites of Oxidized/Hydroxidized Biochar:** Three processes are used to synthesize oxide/hydroxide biochar nanocomposites:
    i. Bioaccumulation of target elements;
    ii. Pre-treatment of biomass with metal salts; and
    iii. Post-pyrolysis incorporation of nanometal oxide.

    The first two methods involved pre-pyrolyzing biomass with metal impregnation, while the 3rd method added oxide/hydroxide nanometal directly to the pyrolyzed biochar [34].
2. **Biochar Coated with Functional Nanoparticles/Functional Nano-Biochar:** To create superior nanomaterial composites capable of eliminating a range of contaminants, biochar may be coated with functional nanoparticles such as graphene and its oxides, chitosan, carbon nanotubes, and layered double hydroxides, among others.
3. **Magnetic Nano-Biochar/Magnetic Biochar Composites:** To make biochar-based magnetic composites, two approaches can be used based on:
    i. Pre-biomass with iron oxides;
    ii. Chemically precipitation of iron oxides onto biochar.

These procedures may coat particles of $CoFeO_4$, Fe_3O_4, α-Fe_2O_3, or nanoscale magnetic iron oxide [35–37] onto the biochar's surface, ensuring the formation of polarity and the availability of iron oxide's catalytic surface for contaminant elimination. The magnetic characteristics of biochar boosted its ability to adsorb and made it simpler to separate from solutions [27]. Composites with improved properties, such as enhanced specific surface area, active site availability, catalytic degradation capacity, pore-size distribution, pore volume, and ease of separation, have successfully removed pollutants with good efficacy. These composites were found to be more effective at adsorbing different pollutants from aqueous solutions, according to numerous studies [38]. Furthermore, these nanocomposites, which comprise a variety of reductive/oxidative, catalytic nanoparticles (such as graphitic C_3N_4 and nanoscale zero-valent iron, etc.), simultaneously adsorb and degrade a number of potentially harmful chemicals [27]. Because of this, nanocomposites made of biochar have a lot of potential for use in many environmental remediation as well as for treating a diverse range of wastes.

7.5 SIGNIFICANCE OF NANO-BIOCHAR IN SOIL-MICROBE INTERACTION: BIOREMEDIATION IN SOIL SYSTEM

7.5.1 PESTICIDES REMEDIATION USING NANO-BIOCHAR

Because of this, nanocomposites made of biochar have a lot of potential for use in many environmental applications as well as for treating a variety of wastes. Every pesticide may be a soil or water pollutant due to its frequently long, continuous days and high utility costs, along with flood or erosion occurrences. It is well known that carbon nanomaterials like nano-biochar have a strong affinity for insecticides. A few ash-C products drastically reduced the phytotoxicity of the herbicide diuron 40 years ago [39]. Numerous research studies revealed that the efficiency of pesticides decreased when combustion residues were present in the soil [23]. Kookana et al. [40] indicated that doubling the diuron injection rate failed to limit weed growth in the presence of 0.5% wheat char in the soil. However, the usage of biochar has recently limited the efficacy of herbicides and weed management. This was due to the biochar's specific surface area (SSA), with higher SSA biochar resulting in lower weed control [41]. Some *in vitro* studies showed that when herbicide treatment rates must be raised for biochar-amended soils, there might be agronomic and/or economic implications, taking into account the farmer's increased input cost of pesticides. Tang et al. [42] observed that adding biochar to soil improved atrazine's ability to control ryegrass weed, and they hypothesized that the recommended dosage for biochar-amended soil (1% w/w) may need to be increased by 3–4 times in order to achieve the required weed control. They also noted that the chemical makeup of the herbicide molecule and its mechanism depended on the amount of biochar's impact on herbicide effectiveness. However, it has been determined that the aging of biochars may eventually result in a decline in sorption capacity, which is crucial for herbicide effectiveness control in biochar-amended soils [43]. The use of biochar for pesticide remediation in pesticide-contaminated ecosystems has the following advantages: (i) it increases soil water retention capacity; (ii) it improves air diffusion conditions in soil; and (iii) it provides an ecological landscape for the best growth of microbes, aiding the microbial population in pesticide degradation through metabolic actions [44]. Concerns about environmental pollution and human exposure from the viewpoint of the ecosystem and human health may be lessened by the increased sorption and decreased management of certain pesticides in the presence of biochar. Second, the increased crop production and decreased organic pesticide residues in crops may result from reduced bioavailability and plant absorption. Although the use of biochar additions to get rid of harmful pesticides and herbicides in agroecosystems seems feasible, more research on pesticide adsorption by biochar is anticipated.

7.5.2 HEAVY METAL REMEDIATION USING NANO-BIOCHAR

One of the most serious and obvious problems in water bodies is pollution, particularly that caused by heavy metals (HMs), a global environmental phenomenon. Additionally, it has been rapidly rising on a worldwide scale [45]. Several anthropogenic activities result in the emission of heavy metals into the environment, including the use of fertilizers and pesticides, irrigation of agricultural land with sewage, mining, the discharge of trash from electronic production, and the recycling and disposal of electronic waste. In the aquatic environment, heavy metals including Cd, Cr, and Ni are recalcitrant and can bioaccumulate and biomagnify in food chains. Due to their toxicity even at low doses, they can gravely impair human health. Therefore, it is vital to remediate these pollutants [44]. By immobilizing the metals into stable forms, soil amendments can lower the bioavailability of metals in soils. Many other kinds of ecologically friendly soil additions, including biochar, have been identified and have attracted a lot of interest recently. Under low oxygen concentrations, inadequate feedstock, such as wood waste, and organic substances are pyrolyzed, such as crop wastes from agriculture, producing biochar due to its sorptive properties. Previous investigations from various places have shown that biochar is excellent at reducing heavy metals (Table 7.1) [42]. By altering the chemical characteristics of the soil, biochar additions may induce positive circumstances for heavy metal immobilization, hence lowering plant absorption of heavy metals [46]. The relevance of nanoremediation has increased recently in terms of improving the environment and preventing, detecting, regulating, and removing pollution, particularly that brought on by chlorinated compounds, hydrocarbons, organic compounds, and heavy metals [47]. Due to their substantial microinterface, many micropores, and substantial surface area, these nanoscale materials contribute to the absorption of heavy metals. Nanomaterials having the potential to have a beneficial impact on the environment are thus quite interesting. In situ, nanomaterial remediation technology and biochar remediation technology, according to

Liu et al. [48] in Scientific Reports (2020), may become more crucial in fostering plant development and improving soil qualities. The usefulness of nano-biochar for promoting plant development and soil microbial variety under excess Cd was therefore evaluated, as well as the suppression of plant growth by Cd. The phytotoxic effects of Cd_{2+} on the germination of plants and the restoration potential of nano-biochar were investigated in a lab experiment utilizing various doses of nano-biochar and Cd_{2+} solution. To further understand the impacts of nano-biochar on the dynamics of the cadmium content and the microbial community in contaminated soil, a pot experiment was also conducted in soil with simulated cadmium contamination. Monitoring Cd phytotoxicity, Cd availability in soil, and abundance in the microbial community allowed researchers to examine the effects of applying nano-biochar to Cd-contaminated soils. The findings showed that adding nano-biochar to Cd-contaminated soil dramatically enhances seedling growth, plant biomass, and the variety of microorganisms. The use of nano-biochar can greatly lower the amount of Cd that is readily accessible in soil and enhance the retention of cadmium ions in soil [49]. The number, variety, and microbial biomass of Actinobacteria and Bacteroidetes, which are helpful for the remediation of polluted soil, were also dramatically increased by the addition of nano-biochar. Nano-biochar is anticipated to be employed extensively in high-concentration cadmium-contaminated fields as a low-cost substance to protect human health and the safety of food. The nanocomposites demonstrated the improved adsorption capacity of copper (Cu (II)) ions with 220.4 mg/g and 185.4 mg/g for lead (Pb (II)) ions [50].

TABLE 7.1 Heavy Metal Removal by Nano-Biochar

Feedstock	Nanomaterials	Heavy Metals	Removal Efficiency	References
Wheat straw	Magnetic nano-biochar	Hg (II)	127.4 mg/g	[51]
Sugarcane bagasse	ZnO-biochar	Cr (VI)	102.66 mg/g	[52]
Wetland reed	Magnetic (Fe) nano-biochar	Hexavalent Cr	58.82 mg/g	[53]
Rape straw	$FeCl_3$ magnetic biochar	Cd (II)	32.74 mg/g	[54]
Cottonseed hulls	Nano MnO_2 biochar	Cu (II)	26.88 mg/g	[55]
Dendro-biochar	Graphitic nano-biochar	Cd (II)	922.27 mg/g	[56]
Red oak	Iron- biochar	As	15.58 mg/g	[57]

7.6 SIGNIFICANCE OF NANO-BIOCHAR IN SOIL MICROBE INTERACTION: IN THE AGRICULTURE ASPECT

7.6.1 PLANT GROWTH

Industrialization, deforestation, overgrazing, and other human activities have dramatically decreased the amount of nutrients in the soil. Lower yield and plant growth as a result of nutrient depletion have a significant negative impact on food security [58]. In an effort to keep soil nutrients in place, chemical fertilizers have been employed excessively. In contrast, nitrogen loss decreased soil bioavailability and prompted the use of artificial fertilizers. Widespread soil pollution and detrimental effects on soil biota were caused by this [59]. In addition to improving soil fertility, permeability, and water-holding capacity, biochar also carries enormous amounts of nutrients due to its high charge density, resulting in an increase in crop output [61].

A crucial ingredient for plant growth is phosphorus (P) [61]. Based on P adsorption, desorption, and transformation between soil solids and solution, phosphate anions (H_2PO_4) are a form of P that plants can absorb from the soil solution. P adsorption and desorption control soil P, altering plant P uptake and usage. Due to its ability to influence the soil, biochar can be used to amend soil. It has been shown to increase P and N availability [62] while lowering P fixation, nitrogen leaching, and N_2O emission. Researchers have found that biochar amendment has a favorable effect on plants. Particularly, biochar is used to promote plant growth by making nutrient delivery simple, holding onto minerals, and offering a variety of options. As a result, biochar enhances plant growth and is crucial for managing soil nitrogen levels for agricultural output [63]. Once more, adding biochar to soil had an impact on properties like density, porosity, particle size distribution, and texture. Because it contains organic material, which raises the pH, cation exchange capacity (CEC), electrical conductivity (EC), and amounts of nutrients like C, N, and P in the soil [64]. This has a positive impact on metal sorption effectiveness. Additionally, it significantly changed the soil's micro-biological and nutritional quality, air/oxygen content, and water holding capacity. Gonzaga et al. [65] revealed that an equal amount of 30 t ha_{-1} of coconut husk biochar added to plant soil caused an increase in plant N and P concentrations of 0.88% and 0.12%, respectively, and a 90%

improvement in maize biomass. As a result, the increase in plant nutrition and plant biomass demonstrates how thoroughly ground coconut husk biochar possesses an extra positive impact on soil. Utilizing biochar decreased the tensile strength of the soil, which decreased the soil compaction [66]. According to several studies, using biochar in conjunction with various contemporary technologies and tailored nanomaterials may lead to superior results.

7.6.2 SOIL REMEDIATION

There are widespread worries about heavy metals (HMs) contamination of soils, which has led to a rise in the demand for soil remediation. To lessen the hazards of HMs in contaminated soils, a lot of work has gone into developing effective remediation techniques, such as thermal treatment, soil washing, excavation of contaminated soils, and phytoremediation [51]. Using biochar, a solid carbonaceous substance, as a remediation method is one that is both economical and environmentally benign. It has been noted that biochar can alter soil redox processes and enhance soil properties, such as facilitating the breakdown of pentachlorophenol and raising the pH of acidic and neutral soils [67]. There is growing interest in using biochar for soil remediation due to its well-established ability to effectively immobilize metals in contaminated soils. HMs polluted areas typically come from a variety of anthropogenic sources, including the burning, mining, and smelting of coal, as well as tanning, textile, and distilling operations.

Although biochar is an eco-friendly way to clean up HMs-affected soils, it is not as frequently used on contaminated sites as it is on agricultural land [68]. When biochar is added to soil, high quantities of minerals (such as Na, Ca, P, Mg, and K) may be released into the soil. On biochar surfaces, the released minerals might create mineral phases and absorb metals from the soil solution. For instance, the production of persistent phosphate minerals results in Pb (II) retention as soil P concentration rises with increased biochar application rates [69]. Because of the complexation between metals and oxygen-containing functional groups in biochar, the release of dissolved organic carbon from biochar and its amendment would raise the soil's organic carbon content (OC), which could decrease metal mobility and bioavailability [70].

Studies revealed that the addition of biochar made from dairy manure to the soils at E-waste recycling facilities showed that biochar could successfully reduce the bioavailability and mobility of HMs. Wood biochar generally enhanced soil conditions and reduced the bioavailability of Cd (II), Zn, and Pb (II), at an annual absorption rate of 5% biochar (w/w) [71].

7.6.3 NANO-BIOCHAR-BASED FERTILIZER

Biochar made from biomass contains high concentrations of micronutrients like iron (Fe), copper (Cu), manganese (Mn), zinc (Zn), and ash, as well as other macronutrients like nitrogen (N), phosphorus (P), magnesium (Mg), potassium (K), calcium (Ca), and sulfur (S) [72]. This makes biochar a high-performance organic fertilizer. Biochar made from animal sources like broiler litter and sewage sludge was found to contain more total P and N than biochar made from wood and green waste like plant biomass [73]. A notable increase in production was seen when biochar was combined with conventional fertilizers, such as urea, while N intake was noticeably reduced. Additionally, the usage of biochar composites containing minerals and organic materials can enhance the benefits of fertilizer use. Due to their ability to retain water and nutrients and their ease of use as nano fertilizers, biochar-derived nanocarbons have attracted considerable interest [74]. It was also reported that the development and productivity of the wheat crop were noticeably boosted by the application of biochar-carbon nanoparticles [75].

7.6.4 BIOCONTROL OF PHYTOPATHOGENS

Global food security is significantly decreased as a result of soil-borne diseases, which result in significant crop output losses of between 10% and 20% [76]. Numerous studies document the influence of fungi and the prevalence of soil-borne plant diseases [77]. These infections affect different plant organs, causing rots, wilts, deformities, spots, chlorosis, discoloration, and other symptoms. Chemical, biological, or resistant strains of pathogens are examples of traditional approaches used to control soil-borne infections [78]. However, the majority of techniques now in use have typically failed to demonstrate sustainable management of infections and their significant financial and environmental implications. Scientific research into adding biochar as a soil additive to capture carbon and boost soil productivity has been advancing quickly in recent years [79]. Additionally, biochar has been successful as a suitable carrier for rhizobacteria that promotes plant growth, enhances crop growth and development, and provides biological control of plant diseases [80]. To increase soil fertility and crop yields, multiple field and laboratory trials with various forms of biochar have recently been carried out. Studies on the use of biochar to prevent diseases brought on by foliar and soil-borne plant pathogens have also been conducted [77, 81].

Although it has previously been shown that biochar amendments help plants become more resistant to disease, the precise processes are yet unknown [82]. The benefits of biochar on a plant's ability to react to biotic stressors can differ and may be either direct or indirect [82]. Biochar may inhibit plant diseases in a number of ways, including improved soil physico-chemical characteristics, increased microbial diversity and abundance, stimulation of systemic plant defense responses, and reduction of pathogen growth, survival, virulence, and activity [83]. In contrast to foliar infections, biochar addition proved more effective when it directly interacted with soil-borne pathogens [84]. These quick results were attributed to enhanced nutrient uptake and digestion, the promotion of beneficial soil bacteria, the beginning of the synthesis of growth-regulating chemicals like elicitors or phytohormones, or the reduction of pathogenic toxin production [85]. A number of the components that make up biochar also include free radicals or offer minerals and volatile organic compounds (VOCs). Applications of biochar to soil have been shown to suppress plant infections by promoting systemic resistance against fungi, such as *F. oxysporum* f. sp. *lycopersici* in tomato plants [83].

The application of biochar effectively boosts plant disease resistance through the control of biochemical processes, physiological responses, and systemic resistance in plants. Thus, biochar amendments can be a dependable strategy for enhancing plant health and guarding against fungal infections that are transmitted through the soil, thereby fostering long-term agro-sustainability [87]. But in the future, it will take more investigation to clarify the mechanisms and pinpoint the metabolites that contribute to induced systemic resistance in plant-pathogen interactions in biochar-amended soil. The global biochar market is expanding, and long-term field studies are essential to advance understanding in this field and make biochar a workable plan of action for controlling plant diseases.

7.6.5 EFFECT OF NANO-BIOCHAR ON SOIL MICROBE COMMUNITY

Microbial populations are necessary for the functioning of the plant soil ecosystem, which includes soil health, carbon fertility, cycling of nutrients, and crop yields. Numerous fixed soil physical parameters have been demonstrated to vary when using biochar [88]. Surface area, bulk density, porosity, and tensile strength are characteristics that often alter the soil biota. The physical and chemical pathways that biochar induces can affect the function and diversity of the soil microbial population [21]. By altering the physical environment of the soil, biochar may have an impact on the profile of the microbe population. However, biochar-induced changes in microbial growth may also change the physical properties of the soil. The abundance, variety, and organization of the microbial community can undoubtedly be influenced by biochar-induced physicochemical changes in the soil, which in turn can affect how plants and microbes interact with each other in the soil [89].

The significance of biochar in the soil changes in pH, and oxidation potential may have an impact on the population structure and activity of microbes [90]. The presence of more biochar in the soil has the potential to affect how quickly mineralizable carbon and nitrogen fertilizers are absorbed, which could alter the behavior of bacterial and fungal communities [91]. The ability of biochar to conveniently supply decaying organic metabolites and nutrients for microbial development is related to its potential to have a favorable impact on microbial activity and community [92]. High surface area and diverse charge biochar materials support microbial growth and nutrient ion sorption. The highly porous structure of biochar can provide essential organic metabolites and minerals required for microbial development as well as a protective microhabitat for soil microbial populations [93]. Improvements in abiotic parameters are being investigated together with a thorough analysis of 6 potential biochar pathways that may determine soil microbial populations [88].

The synthesis of biochar-based systems to restore degraded soils, maintain soil health, and protect plants may benefit greatly from an understanding of how biochar impacts the diversity and functionality of microbial populations (Figure 7.2).

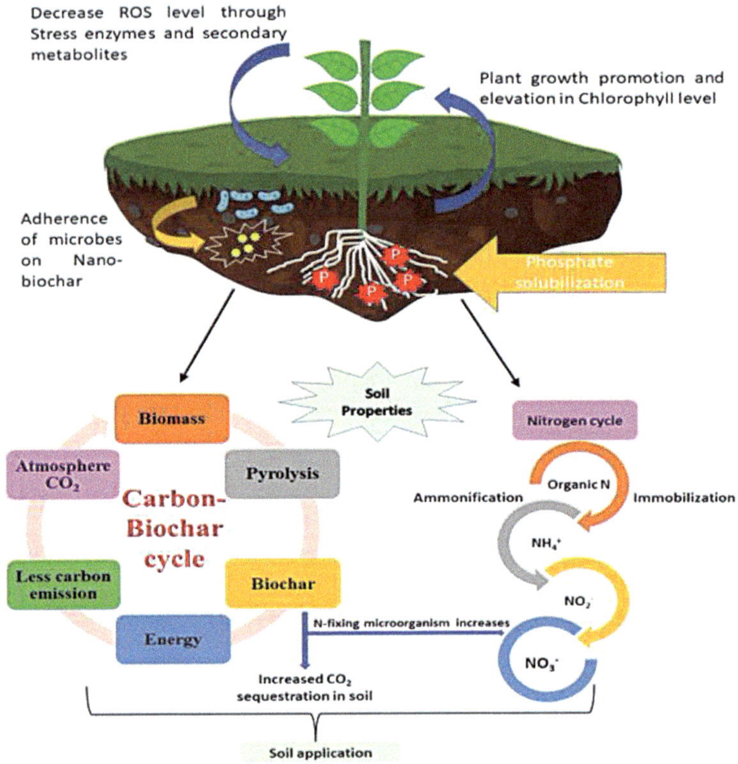

FIGURE 7.2 Significance of nano-biochar in soil microbe interaction and plant growth promotion.

7.6.6 CARBON SEQUESTRATION

Carbon build-up in a stable solid state is referred to as carbon sequestration. Black carbon, also known as biochar, is primarily produced by pyrolyzing woody materials, plant biomass, and straw residues at high temperatures and pressures without oxygen [94]. When feedstock is pyrolyzed, 50% of the carbon in its original biomass accumulates, making it extremely refractory and remaining in the soil for comparatively extended periods of time [95]. Additionally, because it takes millions of years to break down and is very resilient to microbial assault in soil, the process of SOC release into the atmosphere is slowed down over time [7]. In order to increase the concentration of carbon in the soil, crop debris and other organic materials are used in the long-term capture and binding of atmospheric CO_2 into the soil [96]. In general, it describes the direct or indirect conversion of atmospheric CO_2 into a more stable form. While direct C sequestration entails fixing CO_2 into plant biomass through photosynthesis, indirect C sequestration happens when CO_2 is chemically transformed into inorganic substances like calcium carbonate ($CaCO_3$) or magnesium carbonates ($MgCO_3$). All living things contain carbon, and depending on their potential for absorbing and releasing it, they are either referred to as sources or sinks of carbon. Carbon sequestration is crucial because it lessens the greenhouse effect and revives soil activity [97]. By lowering pollutants and preventing global warming, it enhances ecological approaches and promotes sustainable growth. Water affects the soil's ability to store carbon because wet soils are more likely to release carbon than dry soils, which alters the flux of carbon into the atmosphere [98].

Char characteristics are greatly affected by the pyrolysis temperature. While biochar produced at lower pyrolysis temperatures (between 250 and 400°C) is more suited to eliminating inorganic pollutants from the environment, biochar produced at higher pyrolysis temperatures (such as more than 500°C) has a greater capacity to remove organic pollutants [99].

Since ancient times, plants have used the process of photosynthesis to absorb CO_2 that is released by the breakdown of organic matter and the burning of fossil fuels. The carbon contents in biomass are transformed during the pyrolysis of these plant leftovers into stable forms that are bonded to the resultant substance (biochar) and last in soil for generations. It has been determined that building up carbon in biochar could stop the release of CO_2, reducing the amount of CO_2 in the atmosphere and maintaining the balance of carbon in soil and the environment [100].

7.7 FUTURE PERSPECTIVE

In future research, emphasis can be on selecting feedstocks with the right composition, improving production conditions, and adjusting synthesis parameters to create nanocomposites based on biochar that is better able to remove particular pollutants. The mechanisms underlying the nano additive effect of biochar on plant development, the microbial population, and the environment still need further study. While biochar can serve as a nanocarrier for the delivery of insecticides, nucleic acids, and fertilizers to crops and plants, plant nanobionics is another area to investigate the potential of biochar.

7.8 CONCLUSION

Biochar nanocomposites or nano-biochar are utilized to solve agricultural and environmental issues. Nano-biochar can efficiently absorb heavy metals, dyes, medicines, and other organic and inorganic pollutants. Improvements in soil aeration and water holding capacity brought through the addition of biochar may also lower N_2O emissions by reducing the occurrence of anaerobic pockets, which are areas where bacterial-mediated denitrification processes are more likely to take place. Furthermore, there are numerous applications for nano-biochar, including considerably enhancing plant growth and yield, recycling nutrients, pesticide remediation, fertilizer applications, soil amendment, and the removal of pathogenic effects, among others. Synthesis methods for biochar-based nanocomposites and their beneficial impact on enhancing the removal of heavy metal capacity through biochar have been demonstrated by combining the advantages of biochar with nanomaterials. In order to expand the use of biochar and nanotechnology in environmental applications, it has become crucial to impregnate raw biochar with nanomaterials.

KEYWORDS

- **biochar**
- **carbon nanotube**
- **cation exchange capacity**
- **electrical conductivity**
- **nano-biochar**
- **soil-microbe interaction**
- **specific surface area**
- **volatile organic compounds**

REFERENCES

1. Henry, R. K., Yongsheng, Z., & Jun, D. (2006). Municipal solid waste management challenges in developing countries – Kenyan case study. *Waste Management, 26,* 92–100. https://doi.org/10.1016/j.wasman.2005.03.007.
2. Memon, M. A. (2012). Integrated solid waste management based on the 3R approach. In *Sustainable Solid Waste Management* (pp. 133–149). Apple Academic Press.
3. Scarlat, N., Motola, V., & Dallemand, J. F. (2015). Evaluation of energy potential of Municipal Solid Waste from African urban areas. *Renewable and Sustainable Energy Reviews, 50,* 1269–1286.
4. Chopra, R. (2016). Environmental degradation in India: Causes and consequences. *International Journal of Applied Environmental Sciences, 11,* 1593–1601.
5. Portugal-Pereira, J., & Lee, L. (2016). Economic and environmental benefits of waste-to-energy technologies for debris recovery in disaster-hit Northeast Japan. *Journal of Cleaner Production, 112,* 4419–4429. https://doi.org/10.1016/j.jclepro.2015.05.083.
6. Vitorino de Souza Melaré, A., Montenegro González, S., Faceli, K., & Casadei, V. (2017). Technologies and decision support systems to aid solid-waste management: A systematic review. *Waste Management, 59,* 567–584.
7. Lehmann, J., Gaunt, J., & Rondon, M. (2006). Bio-char sequestration in terrestrial ecosystems – A review. *Mitigation and Adaptation Strategies for Global Change, 11,* 403–427.
8. Inyang, M., Gao, B., & Zimmerman, A. (2015). Sorption and cosorption of lead and sulfapyridine on carbon nanotube-modified biochars. *Environmental Science and Pollution Research, 22,* 1868–1876. https://doi.org/10.1007/s11356-014-2740-z.
9. Sohi, S., Lopez-Capel, E., Krull, E., & Bol, R. (2009). Biochar, climate change and soil: A review to guide future research. *CSIRO Land and Water Science Report*, *5*(9), 17–31.
10. Glaser, B., & Birk, J. J. (2012). State of the scientific knowledge on properties and genesis of Anthropogenic Dark Earths in Central Amazonia (terra preta de índio). *Geochimica et Cosmochimica Acta, 82,* 39–51. https://doi.org/10.1016/j.gca.2010.11.029.
11. Lehmann, J. (2007). Bio-energy in the black. *Frontiers in Ecology and the Environment, 5,* 381–387.
12. Chan, K. Y., van Zwieten, L., & Meszaros, I. (2007). Agronomic values of green waste biochar as a soil amendment. *Australian Journal of Soil Research, 45,* 629–634. https://doi.org/10.1071/SR07109.
13. Singh, A., Sengar, R. S., Shahi, U. P., Rajput, V. D., Minkina, T., & Ghazaryan, K. A. (2022). Prominent effects of zinc oxide nanoparticles on roots of rice (*Oryza sativa L.*) grown under salinity stress. *Stresses, 3,* 33–46. https://doi.org/10.3390/stresses3010004.
14. Chan, K. Y., van Zwieten, L., & Meszaros, I. (2008). Using poultry litter biochars as soil amendments. *Australian Journal of Soil Research, 46,* 437–444. https://doi.org/10.1071/SR08036.
15. Headrick, D. R. (2020). *Humans Versus Nature: A Global Environmental History.* Oxford University Press, USA.

16. Zech, W., Haumaier, L., & Hempfling, R. (1990). Ecological aspects of soil organic matter in tropical land use. *Humic Substances in Soil and Crop Sciences: Selected Readings*, 187–202.
17. Tripathi, S., Kapri, S., Datta, A., & Bhattacharyya, S. (2016). Influence of the morphology of carbon nanostructures on the stimulated growth of gram plant. *RSC Advances, 6,* 43864–43873. https://doi.org/10.1039/c6ra01163b.
18. Schulz, H., & Glaser, B. (2012). Effects of biochar compared to organic and inorganic fertilizers on soil quality and plant growth in a greenhouse experiment. *Journal of Plant Nutrition and Soil Science, 175,* 410–422. https://doi.org/10.1002/jpln.201100143.
19. Kavitha, B., Reddy, P. V. L., & Kim, B. (2018). Benefits and limitations of biochar amendment in agricultural soils: A review. *Journal of Environmental Management, 227,* 146–154.
20. Beesley, L., Moreno-Jiménez, E., & Gomez-Eyles, J. L. (2010). Effects of biochar and green waste compost amendments on mobility, bioavailability and toxicity of inorganic and organic contaminants in a multi-element polluted soil. *Environmental Pollution, 158,* 2282–2287. https://doi.org/10.1016/j.envpol.2010.02.003.
21. Lehmann, J., Rillig, M. C., & Thies, J. (2011). Biochar effects on soil biota – A review. *Soil Biology and Biochemistry, 43,* 1812–1836. https://doi.org/10.1016/j.soilbio.2011.04.022.
22. Singh Sekhon, B. (2014). Nanotechnology in agri-food production: An overview. *Nanotechnology for Science and Applications, 7,* 31–53. https://doi.org/10.2147/NSA.S50714.
23. Gao, L., Nie, L., Wang, T., & Liu, L. (2006). Carbon nanotube delivery of the GFP gene into mammalian cells. *ChemBioChem, 7,* 239–242. https://doi.org/10.1002/cbic.200500227.
24. Zhang, Z., Zhu, Z., Shen, B., & Liu, L. (2019). Insights into biochar and hydrochar production and applications: A review. *Energy, 171,* 581–598. https://doi.org/10.1016/j.energy.2019.01.059.
25. Yang, H., Ye, S., Zeng, Z., et al. (2020). Utilization of biochar for resource recovery from water: A review. *Chemical Engineering Journal, 397,* Article 125413. https://doi.org/10.1016/j.cej.2020.125413.
26. Wei, D., Li, B., Huang, H., et al. (2018). Biochar-based functional materials in the purification of agricultural wastewater: Fabrication, application and future research needs. *Chemosphere, 197,* 165–180. https://doi.org/10.1016/j.chemosphere.2018.01.155.
27. Tan, X. F., Liu, Y. G., Gu, Y. L., et al. (2016). Biochar-based nano-composites for the decontamination of wastewater: A review. *Bioresource Technology, 212,* 318–333. https://doi.org/10.1016/j.biortech.2016.04.001.
28. Saxena, M., Maity, S., & Sarkar, S. (2014). Carbon nanoparticles in "biochar" boost wheat (*Triticum aestivum*) plant growth. *RSC Advances, 4,* 39948–39954. https://doi.org/10.1039/c4ra06535b.
29. Jenie, S. N., Kristiani, A., Kustomo, K., Simanungkalit, S., & Mansur, D. (2017). Preparation of nanobiochar as magnetic solid acid catalyst by pyrolysis-carbonization from oil palm empty fruit bunches. In *AIP Conference Proceedings* (Vol. 1904, No. 1). AIP Publishing.

30. Rajput, V. D., Minkina, T., Ahmed, B., Singh, V. K., Mandzhieva, S., Sushkova, S., ... & Wang, B. (2022). Nano-biochar: A novel solution for sustainable agriculture and environmental remediation. *Environmental Research, 210*, 112891.
31. Meyer, S., Glaser, B., & Quicker, P. (2011). Technical, economical, and climate-related aspects of biochar production technologies: A literature review. *Environmental Science & Technology, 45,* 9473–9483. https://doi.org/10.1021/es201691m.
32. Ashfaq, M., Verma, N., & Khan, S. (2017). Carbon nanofibers as a micronutrient carrier in plants: Efficient translocation and controlled release of Cu nanoparticles. *Environmental Science: Nano, 4,* 138–148. https://doi.org/10.1039/c6en00385k.
33. Khodakovskaya, M. V., Kim, B. S., Kim, J. N., et al. (2013). Carbon nanotubes as plant growth regulators: Effects on tomato growth, reproductive system, and soil microbial community. *Small, 9,* 115–123. https://doi.org/10.1002/smll.201201225.
34. Ramadan, M. M., Asran-Amal, & Abd-Elsalam, K. A. (2019). Micro/nano biochar for sustainable plant health: Present status and future prospects. In: *Carbon Nanomaterials for Agri-food and Environmental Applications.* Elsevier, pp. 323–357.
35. Harikishore Kumar Reddy, D., & Lee, S. M. (2014). Magnetic biochar composite: Facile synthesis, characterization, and application for heavy metal removal. *Colloids and Surfaces A: Physicochemical and Engineering Aspects, 454,* 96–103. https://doi.org/10.1016/j.colsurfa.2014.03.105.
36. Baig, S. A., Zhu, J., Muhammad, N., et al. (2014). Effect of synthesis methods on magnetic Kans grass biochar for enhanced As (III, V) adsorption from aqueous solutions. *Biomass and Bioenergy, 71,* 299–310. https://doi.org/10.1016/j.biombioe.2014.09.027.
37. Wang, S., Gao, B., Zimmerman, A. R., et al. (2015). Removal of arsenic by magnetic biochar prepared from pinewood and natural hematite. *Bioresource Technology, 175,* 391–395. https://doi.org/10.1016/j.biortech.2014.10.104.
38. Yao, Y., Gao, B., Inyang, M., et al. (2011). Biochar derived from anaerobically digested sugar beet tailings: Characterization and phosphate removal potential. *Bioresource Technology, 102,* 6273–6278. https://doi.org/10.1016/j.biortech.2011.03.006.
39. Toth, J., & Milham, P. J. (1975). Activated-carbon and ash-carbon effects on the adsorption and phytotoxicity of diuron. *Weed Research, 15*(3), 171–176.
40. Kookana, R. S., Sarmah, A. K., & van Zwieten, L., et al. (2011). Biochar application to soil. Agronomic and environmental benefits and unintended consequences. In: *Advances in Agronomy, 112,* 103–143. https://doi.org/10.1016/B978-0-12-385538-1.00003-2.
41. Graber, E. R., Tsechansky, L., Gerstl, Z., & Lew, B. (2012). High surface area biochar negatively impacts herbicide efficacy. *Plant and Soil, 353,* 95–106. https://doi.org/10.1007/s11104-011-1012-7.
42. Tang, J., Zhu, W., Kookana, R., & Katayama, A. (2013). Characteristics of biochar and its application in remediation of contaminated soil. *Journal of Bioscience and Bioengineering, 116,* 653–659. https://doi.org/10.1016/j.jbiosc.2013.04.002.
43. Khalid, S., Shahid, M., Murtaza, B., et al. (2020). A critical review of different factors governing the fate of pesticides in soil under biochar application. *Science of the Total Environment, 711, 134645.* https://doi.org/10.1016/j.scitotenv.2019.134645.

44. Zhang, H., Yuan, X., Xiong, T., et al. (2020). Bioremediation of co-contaminated soil with heavy metals and pesticides: Influence factors, mechanisms, and evaluation methods. *Chemical Engineering Journal, 398, 125657.*
45. Saxena, G., Purchase, D., Mulla, S. I., et al. (2020). Phytoremediation of heavy metal-contaminated sites: Eco-environmental concerns, field studies, sustainability issues, and prospects. In: *Reviews of Environmental Contamination and Toxicology.* Springer New York LLC, pp. 71–131.
46. Kamran, M., Malik, Z., Parveen, A., et al. (2020). Ameliorative Effects of Biochar on Rapeseed (*Brassica napus* L.) Growth and Heavy Metal Immobilization in Soil Irrigated with Untreated Wastewater. *Journal of Plant Growth Regulation, 39,* 266–281. https://doi.org/10.1007/s00344-019-09980-3.
47. Vázquez-Núñez, E., Molina-Guerrero, C. E., Peña-Castro, J. M., et al. (2020). Use of nanotechnology for the bioremediation of contaminants: A review. *Processes, 8(7), 826.*
48. Liu, W., Li, Y., Feng, Y., et al. (2020). The effectiveness of nano biochar for reducing phytotoxicity and improving soil remediation in cadmium-contaminated soil. *Scientific Reports, 10.* https://doi.org/10.1038/s41598-020-57954-3.
49. Chausali, N., Saxena, J., & Prasad, R. (2021). Nano biochar and biochar-based nanocomposites: Advances and applications. *Journal of Agriculture and Food Research, 5, 100191.* https://doi.org/10.1016/j.jafr.2021.100191.
50. Vishnu, D., Dhandapani, B., Vaishnavi, G., & Preethi, V. (2022). Synthesis of tri-metallic surface engineered nano biochar from Cynodon dactylon residues in a single step – Batch and column studies for the removal of copper and lead ions. *Chemosphere, 286, 131572.* https://doi.org/10.1016/j.chemosphere.2021.131572.
51. Li, R., Zhang, Y., Deng, H., et al. (2020). Removing tetracycline and Hg (II) with ball-milled magnetic nano biochar and its potential on polluted irrigation water reclamation. *Journal of Hazardous Materials, 384, 121095.* https://doi.org/10.1016/j.jhazmat.2019.121095.
52. Gan, C., Liu, Y., Tan, X., et al. (2015). Effect of porous zinc-biochar nanocomposites on Cr (VI) adsorption from aqueous solution. *RSC Advances, 5,* 35107–35115. https://doi.org/10.1039/c5ra04416b.
53. Zhu, S., Huang, X., Wang, D., et al. (2018). Enhanced hexavalent chromium removal performance and stabilization by magnetic iron nanoparticles assisted biochar in aqueous solution: Mechanisms and application potential. *Chemosphere, 207,* 50–59. https://doi.org/10.1016/j.chemosphere.2018.05.046.
54. Li, B., Yang, L., Wang, C. Q., et al. (2017). Adsorption of Cd (II) from aqueous solutions by rape straw biochar derived from different modification processes. *Chemosphere, 175,* 332–340. https://doi.org/10.1016/j.chemosphere.2017.02.061.
55. Zhou, L., Huang, Y., Qiu, W., et al. (2017). Adsorption properties of nano-MnO_2-biochar composites for copper in aqueous solution. *Molecules, 22, 173.* https://doi.org/10.3390/molecules22010173.
56. Ramanayaka, S., Tsang, D. C. W., Hou, D., et al. (2020). Green synthesis of graphitic nano biochar for the removal of emerging contaminants in aqueous media. *Science of the Total Environment, 706, 135725.* https://doi.org/10.1016/j.scitotenv.2019.135725.

57. Bakshi, S., Banik, C., Rathke, S. J., & Laird, D. A. (2018). Arsenic sorption on zero-valent iron-biochar complexes. *Water Research, 137,* 153–163. https://doi.org/10.1016/j.watres.2018.03.021.
58. Ramadan, M. M., Asran-Amal, & Abd-Elsalam, K. A. (2019). Micro/nano biochar for sustainable plant health: Present status and future prospects. In K. A. Abd-Elsalam (Ed.), *Carbon Nanomaterials for Agri-Food and Environmental Applications* (pp. 323–357). Elsevier.
59. Bah, A., Husni, M. H. A., Teh, C. B. S., et al. (2014). Reducing Runoff Loss of Applied Nutrients in Oil Palm Cultivation Using Controlled-Release Fertilizers. *Advances in Agriculture, 2014.* https://doi.org/10.1155/2014/285387.
60. Helal, M. I., Husein, M. E., Walaa, G., & Mostafa, E. D. (2019). Characterization of agricultural residues-based nano biochar and its efficiency in adsorption/desorption of nutrients. *Int. J. Environ., 8*(2), 130–141.
61. Debicka, M., Kocowicz, A., Weber, J., & Jamroz, E. (2016). Organic matter effects on phosphorus sorption in sandy soils. *Archives of Agronomy and Soil Science, 62*(6), 840–855. https://doi.org/10.1080/03650340.2015.1083981.
62. Zeng, Z., Zhang, S., Li, T. Zhao, F., He, Z., Zhao, H., Yang, X., Wang, H., Zhao, J., & Rafiq, M. T. (2013). Sorption of ammonium and phosphate from aqueous solution by biochar derived from phytoremediation plants. *Journal of Zhejiang University. Science. B, 14*(12), 1152–1161. https://doi.org/10.1631/jzus.B1300102.
63. Glaser, B., Lehmann, J., & Zech, W. (2002). Ameliorating physical and chemical properties of highly weathered soils in the tropics with charcoal – A review. *Biology and Fertility of Soils, 35*(4), 219–230.
64. Liang, B., Lehmann, J., & Solomon, D. (2006). Black carbon increases cation exchange capacity in soils. *Soil Science Society of America Journal, 70*(5), 1719–1730. https://doi.org/10.2136/sssaj2005.0383.
65. Gonzaga, M. I. S., Mackowiak, C., & de Almeida, A. Q. (2018). Positive and negative effects of biochar from coconut husks, orange bagasse, and pine wood chips on maize (*Zea mays* L.) growth and nutrition. *Catena, 162*, 414–420. https://doi.org/10.1016/j.catena.2017.10.018.
66. Chan, K. Y., van Zwieten, L., & Meszaros, I. (2008). Using poultry litter biochars as soil amendments. *Australian Journal of Soil Research, 46*(5), 437–444. https://doi.org/10.1071/SR08036.
67. Zhu, X., Chen, B., Zhu, L., & Xing, B. (2017). Effects and mechanisms of biochar-microbe interactions in soil improvement and pollution remediation: A review. *Environmental Pollution, 227*, 98–115.
68. Verma, S., Bhatt, P., & Verma, A. (2021). Microbial technologies for heavy metal remediation: Effect of process conditions and current practices. *Clean Technologies and Environmental Policy, 25, 1485–1507.* https://doi.org/10.1007/s10098-021-02029-8.
69. Cao, X., Ma, L., Liang, Y., et al. (2011). Simultaneous immobilization of lead and atrazine in contaminated soils using dairy-manure biochar. *Environmental Science & Technology, 45*(11), 4884–4889. https://doi.org/10.1021/es103752u.
70. Meng, J., Tao, M., Wang, L., et al. (2018). Changes in heavy metal bioavailability and speciation from a Pb-Zn mining soil amended with biochars from co-pyrolysis of rice

straw and swine manure. *Science of the Total Environment, 633*, 300–307. https://doi.org/10.1016/j.scitotenv.2018.03.199.

71. Lucchini, P., Quilliam, R. S., DeLuca, T. H., et al. (2014). Increased bioavailability of metals in two contrasting agricultural soils treated with waste wood-derived biochar and ash. *Environmental Science and Pollution Research, 21*(5), 3230–3240. https://doi.org/10.1007/s11356-013-2272-y.
72. Hossain, M. K., Strezov, V. V., & Chan, K. Y. (2011). Influence of pyrolysis temperature on production and nutrient properties of wastewater sludge biochar. *Journal of Environmental Management, 92*(1), 223–228. https://doi.org/10.1016/j.jenvman.2010.09.008.
73. Bopp, C., Christl, I., Schulin, R., & Evangelou, M. W. H. (2016). Biochar as a possible long-term soil amendment for phytostabilisation of TE-contaminated soils. *Environmental Science and Pollution Research, 23*(17), 17449–17458. https://doi.org/10.1007/s11356-016-6935-3.
74. Qian, L., Chen, L., Joseph, S., et al. (2014). Biochar compound fertilizer is an option to reach high productivity but low carbon intensity in rice agriculture in China. *Carbon Management, 5*(2), 145–154. https://doi.org/10.1080/17583004.2014.912866.
75. Zhang, P., Huang, P., Xu, X., et al. (2020). Spectroscopic and molecular characterization of biochar-derived dissolved organic matter and the associations with soil microbial responses. *Science of the Total Environment, 708, 134619*. https://doi.org/10.1016/j.scitotenv.2019.134619.
76. Graber, E. R., Frenkel, O., Jaiswal, A. K., & Elad, Y. (2014). How may biochar influence the severity of diseases caused by soilborne pathogens? *Carbon Management, 5*(3), 169–183.
77. Frenkel, O., Jaiswal, A. K., & Elad, Y. (2017). The effect of biochar on plant diseases: What should we learn while designing biochar substrates? *Journal of Environmental Engineering and Landscape Management, 25*, 105–113.
78. Jaiswal, A. K., Elad, Y., Paudel, I., et al. (2017). Linking the belowground microbial composition, diversity, and activity to soilborne disease suppression and growth promotion of tomato amended with biochar. *Scientific Reports, 7*. https://doi.org/10.1038/srep44382.
79. Kalus, K., Koziel, J. A., & Opaliński, S. (2019). A review of biochar properties and their utilization in crop agriculture and livestock production. *Applied Sciences, 9*(3), 3494.
80. Egamberdieva, D., Reckling, M., & Wirth, S. (2017). Biochar-based *Bradyrhizobium* inoculum improves the growth of lupin (*Lupinus angustifolius* L.) under drought stress. *European Journal of Soil Biology, 78*, 38–42. https://doi.org/10.1016/j.ejsobi.2016.11.007.
81. Egamberdieva, D., Ma, H., & Alimov, J. (2020). Response of soybean to hydrochar-based rhizobium inoculation in loamy sandy soil. *Microorganisms, 8*(11), 1–10. https://doi.org/10.3390/microorganisms8111674.
82. Gao, Y., Lu, Y., & Lin, W. (2019). Biochar suppresses the bacterial wilt of tomatoes by improving soil chemical properties and shifting soil microbial community. *Microorganisms, 7*(12), 676. https://doi.org/10.3390/microorganisms7120676.

83. Akhter, A., Hage-Ahmed, K., Soja, G., & Steinkellner, S. (2016). Potential of *Fusarium* wilt-inducing chlamydospores, in vitro behavior in root exudates and physiology of tomato in biochar and compost amended soil. *Plant and Soil, 406*(1–2), 425–440. https://doi.org/10.1007/s11104-016-2948-4.
84. Wiedner, K., & Glaser, B. (2013). Biochar impact on plant resistance to disease. *Biochar Soil Biota*, 49–76.
85. Mehari, Z. H., Elad, Y., & Rav-David, D. (2015). Induced systemic resistance in tomato (*Solanum lycopersicum*) against *Botrytis cinerea* by biochar amendment involves jasmonic acid signaling. *Plant and Soil, 395*(1–2), 31–44. https://doi.org/10.1007/s11104-015-2445-1.
86. Singh, A., et al. (2020). Monitoring soil salinity and recent advances in the mechanism of salinity tolerance in plants. *Biogeosystem Technique, 7*(2), 66–87. https://doi.org/10.13187/bgt.2020.2.66.
87. Alaylar, B., Güllüce, M., & Egamberdieva, D. (2021). Biochar mediated control of soil-borne phytopathogens. *Environmental Sustainability, 4*, 329–334. https://doi.org/10.1007/s42398-021-00187-5.
88. Thies, J. E., Rillig, M. C., & Graber, E. R. (2015). Biochar effects on the abundance, activity and diversity of the soil biota. In *Biochar for Environmental Management* (pp. 327–389). Routledge.
89. Quilliam, R. S., DeLuca, T. H., & Jones, D. L. (2013). Biochar application reduces nodulation but increases nitrogenase activity in clover. *Plant and Soil, 366*(1–2), 83–92. https://doi.org/10.1007/s11104-012-1411-4.
90. McCormack, S. A., Ostle, N., & Bardgett, R. D. (2013). Biochar in bioenergy cropping systems: Impacts on soil faunal communities and linked ecosystem processes. *GCB Bioenergy, 5*(1), 81–95.
91. Muhammad, N., Dai, Z., Xiao, K., et al. (2014). Changes in microbial community structure due to biochars generated from different feedstocks and their relationships with soil chemical properties. *Geoderma, 226–227*, 270–278. https://doi.org/10.1016/j.geoderma.2014.01.023.
92. Mollinedo, J., Schumacher, T. E., & Chintala, R. (2015). Influence of feedstocks and pyrolysis on biochar's capacity to modify soil water retention characteristics. *Journal of Analytical and Applied Pyrolysis, 114*, 100–108. https://doi.org/10.1016/j.jaap.2015.05.006.
93. Anderson, C. R., Condron, L. M., Clough, T. J., et al. (2011). Biochar induced soil microbial community change: Implications for biogeochemical cycling of carbon, nitrogen, and phosphorus. *Pedobiologia, 54*(6), 309–320. https://doi.org/10.1016/j.pedobi.2011.07.005.
94. Dungait, J. A. J., Hopkins, D. W., Gregory, A. S., & Whitmore, A. P. (2012). Soil organic matter turnover is governed by accessibility, not recalcitrance. *Global Change Biology, 18*(6), 1781–1796.
95. Feola Conz, R., Abbruzzini, T. F., de Andrade, C. A., et al. (2017). Effect of pyrolysis temperature and feedstock type on agricultural properties and stability of biochars. *Agricultural Sciences, 8*, 914–933. https://doi.org/10.4236/as.2017.89067.
96. Izaurralde, R. C., Rosenberg, N. J., & Lal, R. (2001). Mitigation of climatic change by soil carbon sequestration: Issues of science, monitoring, and degraded lands. *Science, 304*(5677), 1623–1627.

97. Lal, R. (2004). Soil carbon sequestration impacts on global climate change and food security. *Science, 304*(5677), 1623–1627.
98. Sofi, J. A., Lone, A. H., Ganie, M. A., et al. (2016). Soil microbiological activity and carbon dynamics in the current climate change scenarios: A review. *Pedosphere, 26*(5), 577–591.
99. Cheng, C. H., Lehmann, J., & Engelhard, M. H. (2008). Natural oxidation of black carbon in soils: Changes in molecular form and surface charge along a climosequence. *Geochimica et Cosmochimica Acta, 72*(7), 1598–1610. https://doi.org/10.1016/j.gca.2008.01.010.
100. Smith, P. (2016). Soil carbon sequestration and biochar as negative emission technologies. *Global Change Biology, 22*(4), 1315–1324. https://doi.org/10.1111/gcb.13178.
101. Rajput, V. D., Minkina, T., Kumari, A., Harish, Singh, V. K., Verma, K. K., Mandzhieva, S., Sushkova, S., Srivastava, S., & Keswani, C. (2021). Coping with the challenges of abiotic stress in plants: New dimensions in the field application of nanoparticles. *Plants, 10*(6), Article 1221. https://doi.org/10.3390/plants10061221.
102. Rajput, V., Minkina, T., Sushkova, S., Behal, A., Maksimov, A., Blicharska, E., Ghazaryan, K., Movsesyan, H., & Barsova, N. (2019). ZnO and CuO nanoparticles: A threat to soil organisms, plants, and human health. *Environmental Geochemistry and Health, 42*(1), 147–158. https://doi.org/10.1007/s10653-019-00317-3.
103. Rajput, V. D., Minkina, T., Sushkova, S., Tsitsuashvili, V., Mandzhieva, S., Gorovtsov, A., Nevidomskyaya, D., & Gromakova, N. (2017). Effect of nanoparticles on crops and soil microbial communities. *Journal of Soils and Sediments, 18*(6), 2179–2187. https://doi.org/10.1007/s11368-017-1793-2.
104. Rajput, V., Minkina, T., Fedorenko, A., Sushkova, S., Mandzhieva, S., Lysenko, V., Duplii, N., Fedorenko, G., Dvadnenko, K., & Ghazaryan, K. (2018). Toxicity of copper oxide nanoparticles on spring barley (*Hordeum sativum distichum*). *Science of The Total Environment, 645*, 1103–1113. https://doi.org/10.1016/j.scitotenv.2018.07.211.
105. Rajput, V. D., Kumari, A., Minkina, T., Barakhov, A., Singh, S., Mandzhieva, S. S., Sushkova, S., Ranjan, A., Rajput, P., & Garg, M. C. (2022). A practical evaluation of the integrated role of biochar and nanomaterials in soil remediation processes. *Environmental Geochemistry and Health, 45, 9435–9449*. https://doi.org/10.1007/s10653-022-01375-w.
106. Rajput, V. D., Faizan, M., Upadhyay, S. K., Kumari, A., Ranjan, A., Sushkova, S., Chauhan, P. K., Mahmud, A. A., Burachevskaya, M., Chaplygin, V., & Deryabkina, I. (2022). Influence of nanoparticles on the plant rhizosphere microbiome. In T. J. Pandey, & A. N. Singh (Eds.), *Sustainable Plant Nutrition in a Changing World* (pp. 83–102). https://doi.org/10.1007/978-3-030-97389-6_4.
107. Rajput, V. D., Singh, A., Minkina, T., Rawat, S., Mandzhieva, S., Sushkova, S., Shuvaeva, V., Nazarenko, O., Rajput, P., Komariah, Singh, A. K., Rao, M., & Upadhyay, S. K. (2021). Nano-enabled products: Challenges and opportunities for sustainable agriculture. *Plants, 10*(12), Article 2727. https://doi.org/10.3390/plants10122727.
108. Singh, A., Singh, A. K., Rawat, S., Pal, N., Rajput, V. D., Minkina, T., Sharma, R., Singh, N. P., & Tripathi, J. N. (2022). Satellite-based quantification of methane

emissions from wetlands and rice paddies ecosystems in north and northeast India. *Hydrobiology, 1*(3), 317–330. https://doi.org/10.3390/hydrobiology1030023.

109. Singh, A., Sengar, R. S., Rajput, V. D., Minkina, T., & Singh, R. K. (2022). Zinc oxide nanoparticles improve salt tolerance in rice seedlings by improving physiological and biochemical indices. *Agriculture, 12*(7), Article 1014. https://doi.org/10.3390/agriculture12071014.
110. Singh, A., Sengar, R. S., Sharma, R., Rajput, P. (2021). Nano-priming technology for sustainable agriculture. *Biogeosystem Technique, 8*(2), 79–92. https://doi.org/10.13187/bgt.2021.2.79.
111. Singh, A., Mehrotra, R., Rajput, V. D., Dmitriev, P., Singh, A. K., Kumar, P., Tomar, R. S., Singh, O., & Singh, A. K. (2022). Geoinformatics, artificial intelligence, sensor technology, big data. In D. P. Mishra (Ed.), *Sustainable Agriculture Systems and Technologies* (pp. 295–313). https://doi.org/10.1002/9781119808565.ch14.
112. Singh, A., Rajput, V. D., Rawat, S., Sharma, R., Kumar Singh, A., Kumar Singh, A., Tomar, R. (2022). Geoinformatics and nanotechnological approaches for coping up abiotic and biotic stress in crop plants. In D. P. Mishra (Ed.), *Sustainable Agriculture Systems and Technologies* (pp. 337–359). https://doi.org/10.1002/9781119808565.ch17.
113. Singh, A., Rajput, V. D., Rawat, S., Sharma, R., Kumar Singh, A., Kumar Singh, A., Tomar, R. S. (2021). Emerging tools for sustainable agriculture and food security: An introduction. In D. P. Mishra (Ed.), *Emerging Tools for Sustainable Agriculture and Food Security* (Vol. 1, pp. 3–16). Deepika.
114. Rajput, P., Rajput, V. D., Singh, A., Sharma, R., Rawat, S., Tomar, R. S., Singh, A., & Minkina, T. (2021). Impact of emerging tools for solving 'hidden hunger' in humans. In D. P. Mishra (Ed.), *Emerging Tools for Sustainable Agriculture and Food Security* (Vol. 1, pp. 17–30). Deepika.
115. Rajput, V. D., Minkina, T., Kumari, A., Shende, S. S., Ranjan, A., Faizan, M., Barakvov, A., Gromovik, A., Gorbunova, N., Rajput, P., Singh, A., Khabirov, I., Nazarenko, O., Sushkova, S., & Kizilkaya, R. (2022). A review on nanobioremediation approaches for restoration of contaminated soil. *Eurasian Journal of Soil Science (EJSS), 11*(1), 43–60. https://doi.org/10.18393/ejss.990605.
116. Rajput, V. D., Minkina, T., Feizi, M., Kumari, A., Khan, M., Mandzhieva, S., Sushkova, S., El-Ramady, H., Verma, K. K., Singh, A., Hullebusch, E. D. van, Singh, R. K., Jatav, H. S., & Choudhary, R. (2021). Effects of silicon and silicon-based nanoparticles on rhizosphere microbiome, plant stress and growth. *Biology, 10*(8), Article 791. https://doi.org/10.3390/biology10080791.

CHAPTER 8

Role of Nanoparticles and Nano-Biochar in Improving Intrinsic Soil Characteristics

FAKHER ABBAS,[1] MAHMOOD UL HASSAN,[2] JIANJUN DU,[1] ALEEZA,[3] RIDA BULBUL,[4] and RUQIN FAN[1]

[1]*Guangdong Engineering and Technology Center for Environmental Pollution Prevention and Control in Agricultural Producing Areas, College of Environmental Science and Engineering, Zhongkai University of Agriculture and Engineering, Guangzhou, China*

[2]*College of Grassland Agriculture, Northwest A&F University, Yangling, China*

[3]*Institute of Soil and Environmental Sciences, PMAS-Arid Agriculture University Rawalpindi, Pakistan*

[4]*Institute of Horticulture, PMAS-Arid Agriculture University Rawalpindi, Pakistan*

ABSTRACT

Recently, the demand for nanoparticles and nano-biochar has surged in the field of agriculture and the environment due to their several benefits with limited harmful effects. Nanotechnology has evolved dramatically in recent decades, as evidenced by a 25-fold rise in consumption between 2005 and 2010. Nano-biochar has demonstrated several benefits over conventional biochar, so it is preferred due to the convenience and benefits

Nano-Biochar and Nanoparticles for Soil Health Improvement.
Vishnu D. Rajput and Tatiana M. Minkina (Eds.)

it offers. Generally, nanomaterials are being used for plant growth, as a soil amendment, and to deal with environmental hazards. Despite a few limitations, there are plenty of benefits of using nanoparticles and nano-biochar, indicating their suitability for agriculture. Along with plants, these materials severely impact soil medium: they affect the physical, chemical, and biological characteristics of the soil. Among the physical properties, they affect soil aggregate stability directly by regulating the bonding of soil particles and indirectly by regulating the soil pores-to-soil particles ratio. Overall, they are used as amendments for soil conservation in soil physics. Nanomaterials influence clay (kaolinite) percentage in the soils, which works as a bonding agent for soil aggregates and eventually improves soil aggregation properties. These nanomaterials increase the soil water infiltration rate by improving the number of soil voids. Among the chemical characteristics of soils, these materials ensure the soil nutrient availability, sorption, and mobilization, hence facilitating the plant. The soil pH and cation exchange capacity (CEC) are two of the most important chemical properties positively influenced by nanomaterials; the alteration in pH is because of nutrient (cations) exchange property and root exudate. The impact of nanomaterials on CEC is significant as it controls nutrient availability and absorbs hazardous materials from the soil. Nanomaterials also improve soil organic matter, as reported by several studies. Soil organic matter plays an important role due to its biologically active organic carbon. SOM is associated with many other soil properties. Hence, the alteration in SOM could cause positive changes in other soil properties, including aggregation and nutrient cycling. Though the nanoparticles and nano-biochar exhibit some limitations, i.e., they sometimes act as a sponge for hazardous material, affect the microbial community, and environmental hazards due to their excessive use, however, these negative effects depend on soil type, duration of exposure, and concentration of nanomaterials (Table 8.1). The chapter has focused on the potential benefits of nanomaterials and nano-biochar for agriculture.

8.1 INTRODUCTION

Nanotechnology has evolved dramatically in recent decades, as evidenced by a 25 times rise in consumption between 2005 and 2010 [1]. Usually, nanomaterials are utilized for plant growth, as a soil amendment, and to address environmental issues. Despite some limitations, nanoparticles and nano-biochar offer numerous advantages, suggesting their appropriateness for agriculture. Along with plants, these materials severely impact soil medium and various soil intrinsic properties. The NMs can be used for several agricultural practices, hence demonstrated to have several advantages over simple mineral particles [2]. Because NM can move through the cellular surface of many different kinds of humans and prokaryotes as prospective biochemical carriers, they have attracted the attention of phyto-scientists [3]. Additionally, due to the fast pyrolysis processing of biomass, along with the biochar biofuels and syngas are also produced. As a result, creating BC or NM nanoparticles may achieve four major aims, namely the elimination of pollutants, the management of waste, carbon sequestration, and power generation [4]. Nanomaterials can affect the percentage of clay (kaolinite) in soils, acting as a bonding agent for soil aggregates and ultimately enhancing soil aggregation properties. These nanomaterials can boost the soil water infiltration rate by improving the number of soil voids [5]. The chapter has focused on the potential benefits of nanomaterials and NB for agriculture.

8.2 ROLE OF NANOPARTICLES IN AGRICULTURE

Nanoparticles (NP) are currently the focus of many researchers in the agricultural sector, particularly in arid and semiarid regions. Typically, the nanoparticle size varies between 1 and 100 nanometers. Nanoparticles are being developed and studied to be used as additives or fillers to achieve a wide range of effects. Due to the significant growth in nanoparticle manufacturing and consumer usage, which has increased nanoparticle exposure and release into the environment, the behavior of nanoparticles in the environment has attracted substantial attention [6]. Nanoparticles form in terrestrial environments either as a result of natural phenomena (e.g., volcanic eruptions, soot from forest fires, fine sand and dust, and biological matter) or as a result of anthropogenic processes, particularly combustion sources (welding, diesel engine exhaust, and coal fires) [7]. In the last decade, synthetic materials have gained prominence in several disciplines, such as medicine, industry, and environmental engineering. They can be categorized based on their morphology [8]. Various NPs can be used for different agricultural products [9].

Chemically generated NPs in the size range of 1–100 nm are often quite helpful. The chemical methods are used in a planned and systematic way to make NPs. Intriguingly, the behaviors of NPs and their bulk (non-NP) counterparts differ dramatically [10]. Carbon nanotubes (CNTs), carbon dots, epoxy resin-coated CNTs, polymer-coated Ag, supermagnetic iron oxide nanoparticles (SPION), mesoporous silica particles, catalytic metals, metal oxides, quantum dots, dendrimers, nanofilms, nanofibers, and composite nanoparticles comprise NPs in a broad sense [11]. In general, the particle size of NPs affects their characteristics; for instance, nanoscale ZnO has a different reaction rate, adsorption capacity, and redox state than bulk ZnO. Similarly, the transition temperature of ferromagnetic particles ($MnFe_2O_4$, $MgFe_2O_4$, etc.) changes significantly with particle size [12]. Therefore, significant efforts have been made to modify the physicochemical parameters (shape, size, and surface charge) of nanoparticles (NPs) to improve their reactivity, strength, and electrical properties [13] (Figure 8.1).

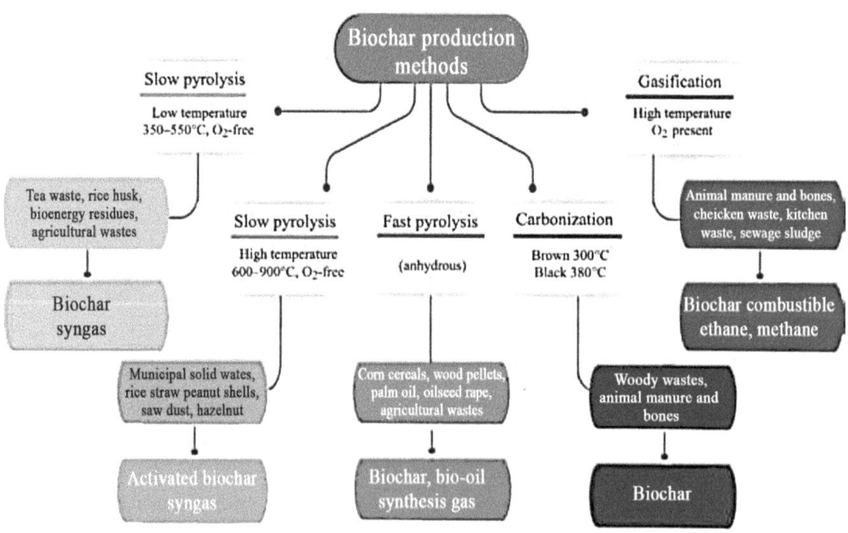

FIGURE 8.1 Flowchart illustration of biochar production approaches.

Source: Reprinted with permission from Ref. [86]. Copyright © 2020 Elsevier Inc.

To assess the eco-toxicity of NPs, it is necessary to understand their agglomeration/aggregation, dissolving rate, surface charge, and surface chemistry. In order to control the biology and fertility of agricultural soils, investigations are also necessary to evaluate the detrimental effects of additional nanoparticles (NPs) on soil physicochemical properties [14]. Another area of study and development relates to how the challenge with NPs alters the expression of genes in plants and microbes that are essential for important cellular functions [15]. However, there are still gaps in our understanding of the potential interactions between soil and nanomaterials that must be filled to accurately determine the effects produced throughout the entire life cycle of each type of nanomaterial.

8.3 ROLE OF NANO-BIOCHAR (NB) IN AGRICULTURE

Adding NB-based fertilizer allowed for changes in the network topology structure, the keystone taxa (algae and fungi), changed pore structure, increased soil productivity, controlled soil pH, EC, and CEC reduced soil-based abiotic stressors are all examples of soil-based processes [16]. All the advantages suggest the possibility that the most recent composites (NB) may be another innovative and creative material that could inexpensively and effectively deal with several key soil issues. Possible methods for producing the NB include progressive thermal combustion [17]. Recent research has also produced a usable nanocomposite by impregnating waste with functional nanomaterials prior to combustion. When Biochar (BC) undergoes pyrolysis and different processes, they are converted into nanometer (nm) sized particles, designated as nano-biochar (Figure 8.2). Recent research produced NB with a dimension of less than 5 nm [18]. The morphological and physiochemical characteristics of BC and NB vary from one another. The mineral contents, aromatic nature, pH, ability to interchange cations, surface area, and porous properties may all differ [19]. NB may also exist in dissolved and hydrophobic stages. In general, the carbon content in the accessible concentration of farmyard waste biochar varies from 14% (100°C) to 0.2% (700°C), with the soluble carbon content being boosted with shifting the pH towards alkaline and decreased with increasing combustion heat [20]. To fully understand the efficient resource utilization of BC, it is necessary to look at how NB affects soil improvement and plant development.

Furthermore, while the characteristics of NB are distinct from BC, its effects on soil health and plant development are presumably different [21]. Among the chemical characteristics of soils, these materials help ensure the availability of soil nutrients, sorption, and mobilization, thus

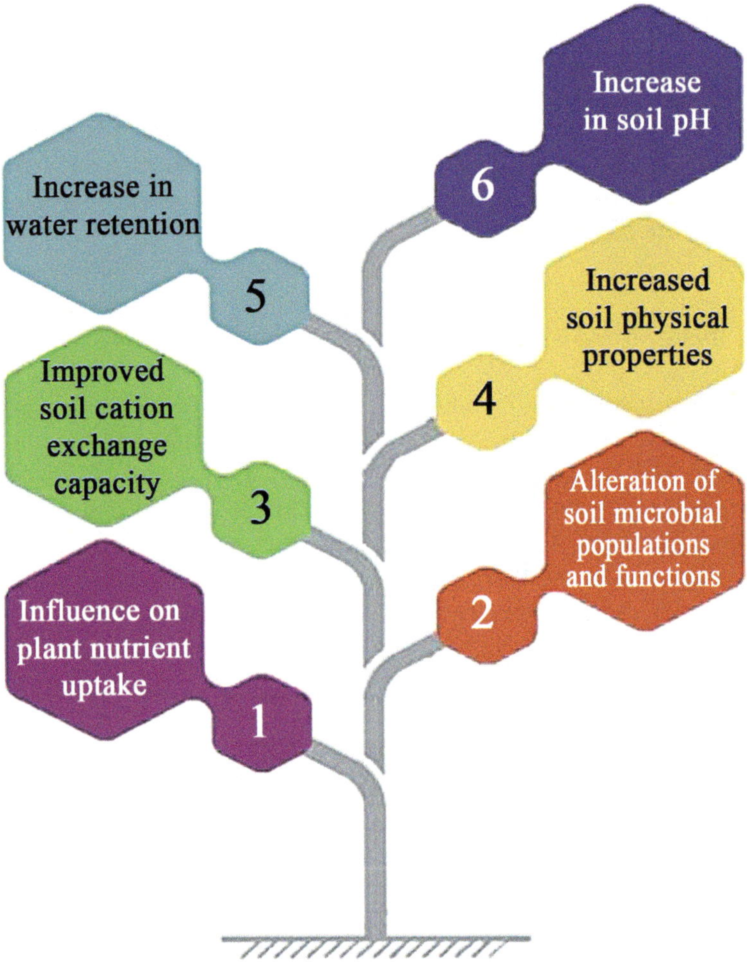

FIGURE 8.2 The mechanism of biochar as a soil conditioner for plant growth includes influence on plant nutrient uptake, alteration of soil microbial populations and functions, improved soil cation exchange capacity, increased soil physical properties, increased water retention, and increased soil pH.

Source: Reprinted with permission from Ref. [86]. Copyright © 2020 Elsevier Inc.

aiding the plants. The impact of nanomaterials on CEC is significant as it controls nutrient availability and absorbs hazardous materials from the soil [22]. Nanomaterials also improve soil organic matter, as reported by several studies. Soil organic matter is important due to its biologically active organic carbon (Figure 8.3). SOM is associated with many other soil properties as well. NB enhances the productivity of organic carbon towards intrinsic soil parameters while decreasing disintegration. As a result, carbon dioxide levels are greatly decreased. Other impacts of raising organic carbon include supporting soil aggregate stability, boosting the soil's capacity to store moisture, enhancing the microbial count, and enhancing the absorption of soil minerals. Enhancing the quality of nano-biochar in the soil favors environmental impact [23]. Nano-biochar has distinct physicochemical features that may be attributed to its production from various plant sources and heating breakdown environments (Figure 8.1). A range of soils may benefit from this substance's various absorption qualities. The synthesis of nano-biochar is a practical, eco-sustainable technique that may enhance the utilization of nutrient sources and water utilization. Additionally, it may be utilized as a foundation for macro- and micronutrients, allowing the minerals to fully process the fragrant fertilizers [24].

Nano-biochar has a variety of characteristics that may boost plant development in salty soil. An additional benefit of Nano-biochar is its capacity to clean the soil of various organic and inorganic pollutants. By using these nanomaterials, it is believed that productive soils will not be at risk of deterioration or degradation, and the output of agricultural goods per unit area would rise [25]. Nano-biochar is a substance rich in carbon that is created during the decomposition and combustion of feedstock. The presence or lack of oxygen distinguishes combustion from thermal breakdown. By using heat and oxygen during the combustion process, biomass may be transformed into gases, silica, and other liquid inorganic components. In the absence of oxygen, the material undergoes thermochemical activation without the production of carbon dioxide gas. In this instance, carbon manifests as biochar, a solid. The thermochemical decomposition also results in the production of liquid bio-oil and syngas, consisting of characteristics of the carbon sequestration technique.

8.4 EFFECT OF NANOPARTICLES AND NANO-BIOCHAR ON SOIL

Most of the nanoparticles end up in the soil. They provide several benefits to the soil and plant environment. The behavior of NPs and NB and the estimation of their effect in the rhizosphere are extremely pertinent concerns. NPs that are introduced artificially in the soil demonstrate

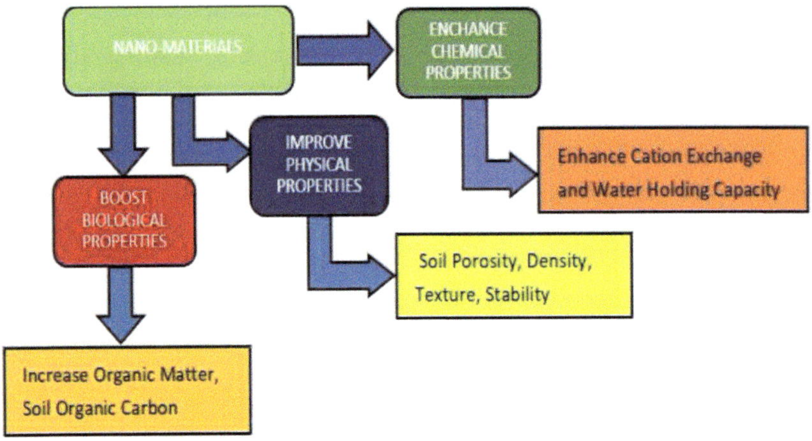

FIGURE 8.3 The flow chart illustrates the role of nanoparticles (nano-biochar and nanoparticles) on soil physical properties (soil porosity, texture, stability), chemical properties (cation exchange capacity, hydraulic conductivity, and water holding capacity), and biological properties (organic matter, humic substance, and soil organic carbon).

different effects than natural NPs. Ben-Moshe et al. [26] revealed the impact of nanoparticles on several microscopic soil characteristics. The effect of nanoparticles on soil relies on their content, the type of soil, and the soil's enzymatic activity. It is essential to investigate the existence of nanoparticles in soil and their effect on the biodiversity of soil. Soil factors such as pH, texture, structure, and organic matter content impact the soil microbial population and the toxicity of contaminants to microorganisms [27]. Nanoparticles may impact the mobility of soil contaminants. Therefore, it is necessary to compare the toxicity of NPs in diverse soil types. The soil modified with digestate and fly ash decreased the bioavailability of the contaminant [28].

8.4.1 HOW NANOMATERIALS INFLUENCE SOIL AGGREGATE STABILITY?

The physical structural differences between soil and biochar cause variations in strength, hydraulic conductivity, and gas transport. The survival of soil organisms may also be impacted by these circumstances. The type and properties of the nano-biochar being used determine all of these consequences (raw material and heat) [29]. The addition of biochar reduces soil strength when its tensile strength is greater than that of the biochar. In a study on Alf sols soils, it was found that adding nano-biochar decreased the tensile stress from 4.64 to 31 kPa. Root development is significantly affected by electromechanical impedance reduction. At this TS, the roots can easily penetrate, and the plant can readily produce roots, although this is just one factor among all the factors affecting it.

Nanomaterials have gained significant scientific attention recently as an alternative for soil stabilization because they may actively interact with increasing clay soil particles [30]. They presented the findings of several tests for unconfined compressive strength. As a result of introducing Modified montmorillonite nano clay, the soil's unconfined compressive strength was significantly improved. The result demonstrated that the soil's liquid limit and plasticity index increased. It may be assumed that Fe-(hydro) oxides promote aggregate formation at greater size scales, improving soil macro aggregate stability and increasing soil porosity. Aggregation is the feature of NPs wherein they are closely bound together by their core. On the other hand, agglomeration refers to the formation of weak intra-NP cementing [31]. Typically, agglomeration occurs when the attraction's kinetic energy surpasses the repulsive energy. However, when connected by their cores, NPs resemble particles that gradually settle down due to gravity. The size variation of these particle flocks is mostly determined by two factors:

i. The concentration of NP in the medium; and
ii. The size of each NP.

Iron oxide nanoparticles quickly agglomerate as their concentrations grow, and these aggregates are extremely persistent [32]. At greater concentrations, Zn-oxide NPs showed various forms and sizes. Moerz et al. [33] showed that when Au-NPs were put into a solution of hemoglobin, the NPs tended to aggregate. When assessing the mechanical quality of soil, macroaggregates in the soil are crucial markers. After being submerged in water, despite disruptions, soil macroaggregates (MA) ensure the soil structure is maintained. Biochar improved the MA concentrations in various soil layers compared to CK, and several studies also found comparable findings [34]. The aggregate contents of the soil's water table layer were all higher in HBC, NBC, and HNBC than in BC. The MA concentration of HBC was greatest in the 0–15 cm soil layer, and it was 1.59–1.96 times and 1.45–1.80 times greater than that of CK and

BC, respectively. These findings imply that HBC significantly increased the amount of soil macroaggregates.

Perhaps because acidification enhanced the abundance of carboxy and made it easier for biochar to combine with soil particles to form soil aggregates, NB impacts soil aggregation [35]. With a greater number of sodium ions, the disruption of aggregates is greatly distributed as a result of the sodium ions' heavy adsorption. Additionally, the MA contents and soil aggregate stability are both dramatically decreased by a lack of soil moisture. Particularly, soil modified with HBC had the biggest decreases in salt content [36] and the greatest improvements in WHC; as a consequence, the HBC treatment had more soil that was water-stable macro-aggregate than other treatments. Soil microaggregates decreased by up to 14 while macroaggregates increased by about 36%. The addition of NB increased the proportion of macroaggregate to the greatest extent, demonstrating the ability of NB to enhance soil structural stability. The most noticeable improvements in SWSA were seen at mass levels of NB up to 1%. As a consequence of multiple substances adhering to NB, the outer layer of the NB is reduced on a molecular level and acts as a stabilizing substance for the development and persistence of structural stability. Particularly in sandy loam soil, the NB improved soil macroaggregates and soil aggregate stability [37].

Based on the amount of biochar fractions added, the pores could either reduce or grow as a result of the addition of biochar, which in turn could change the biochar's permeability. In comparison to the control soil, the porosity of soil amended with NB measurements was reduced. Obia et al. [38] hypothesized that the BC, which would be distinguished by a porous framework with numerous pores of various sizes at the interface, might improve the functional void volume of soil by altering the pore space amongst fractions and replenishing the pores at the surface. Since the mean size of nano biochar's voids is smaller than the particle size of soil, there is no space available for soil structure formation. By obstructing the space between certain soil particles, NB could only have an impact on the soil's porous structure [39]. Since there is a flow rate of water on the soil surface, porous homogeneity brought on by the alignments of soil fractions would change the immediate moisture levels of the soil layer with time.

The primary mechanism through which the NB influenced the rearrangement of soil micropores was the accumulation or closing in of micro-pores of soil; this was immediately represented by the shift in pore characteristics. High NB amounts in the soil led to a considerable reduction in the steady permeability, which in turn led to a reduction in the total combined infiltration rate of the soil. This potentially led to the 'pseudo immersion' phenomenon of the soil below the mixed layer. The NB treatment caused a shift in the fundamental soil stability, which in turn had an effect on the behavior of the soil. The higher contact degree of the inversion layer caused by NB may cause an alteration in the permeability as well as a preferred hydrological cycle [40]. NB decreased the fractional link across porosity and sealed the tiny inter-particle voids. Nevertheless, the rate of NB was larger than 1% in the majority of the trials, and the particle size was > 1 mm level. The NB application had an effect on the water permeability even when the administration rate was changed; in this respect, biochar is comparable. The influence that NB showed on the flow of water at the nano level was supplemented by the effect of NB. When the NB content of the soil increased, the accumulated water infiltration decreased by up to 13%.

8.4.2 EFFECT OF NANOMATERIALS ON SOIL TEXTURE AND MINERALOGY (CLAY%)

The inorganic minerals found in the pedosphere and the extent of weathering are of concern to soil mineralogy. On the other hand, the proportion of silt, sand, and clay-sized particles in the mineral section of the soil is referred to as its texture. When nanomaterials were added, the liquid and plastic limits, dry unit weight, CBR, and unconfined compressive strength increased, whereas the plasticity index and ideal water content decreased. These modifications enhanced the soil behavior. Additionally, the toxicological effects of the Ag_{10}–NH_2 may have been lessened due to the prevention of oxidation caused by the high quantity of natural organic matter produced from the land uses of the clay and loam regions. At this point, negatively charged organic matter began to take over the surface characteristics of positively charged Ag_{10}–NH_2, increasing steric stability and decreasing bioavailability.

8.4.3 HOW NANOMATERIALS IMPACT AND SOIL HYDRAULIC CONDUCTIVITY AND BULK DENSITY?

Hydraulic conductivity is the capacity of a porous media (such as soil) to transfer water under saturated or nearly saturated circumstances, according to scientific terminology. Soil amendment through biochar has the potential to affect the physical and chemical characteristics of the soil, improving soil fertility [41]. Small-sized biochar particles alter the soil's pore structure and have an impact on the porosity, fractal dimension, stability, and degree of anisotropy of soil aggregates. Nano-biochar (NB) enhances smaller soil voids. According to [42], the carbon nanoparticle size in this study demonstrated good adsorption and a large specific surface area, and the application of nano-biochar (NB) boosted the stability of soil aggregates. As a result, the impact of NBC's pore dispersion on soil water movement might be disregarded. When nanomaterials were added, the liquid and plastic limits, dry unit weight, and unconfined compressive strength increased, whereas the plasticity index and ideal water content decreased. These modifications enhanced the soil's behavior. The application of a particular kind of nanomaterial, such as nano clay, does not significantly enhance soil. Excessive nano-clay content, however, has a negative impact. In terms of improvement, nano-copper was more effective than nano-alumina. The lower shrinkage and expansive strains are a direct result of the rise in dry density [43].

Studying the penetration of water into the soil in saline-alkaline land is significant because the transport of salt in the soil is easily influenced by the soil's water content. In comparison to CK, applying biochar enhanced the soil-wetting peak's movement for the same absorption period. The nano-biochar needed longer absorption periods than BC to cover the same wetting distance. NB increased soil permeability (up to 8,000 min) and helped with soil moisture retention [44]. Findings revealed that the inclusion of NB improves moisture retention in the soils and increases permeability. By doing so, the penetration rates and soil hydraulic conductivity increased. Thirdly, the technique of biochar alteration had an impact on the soil's ability to retain soil water. With high pH, the biochar's porosity readily increased. Changing the biochar size to the nano-level improved the soil's wettability and neutrality and enabled it to store water [45] (Figure 8.4).

In contrast to acidity, particle diameter adjustment (nanoparticles) lowered the oxygen-containing functional groups on the biochar to lessen its water permeability and polarization. Additionally, the biochar's previous polymeric configuration and fragment size reduction to the nanoscale disrupted the biochar's original highly porous structure, enabling water to flow through more easily, obstructing soil pore structure, and decreasing

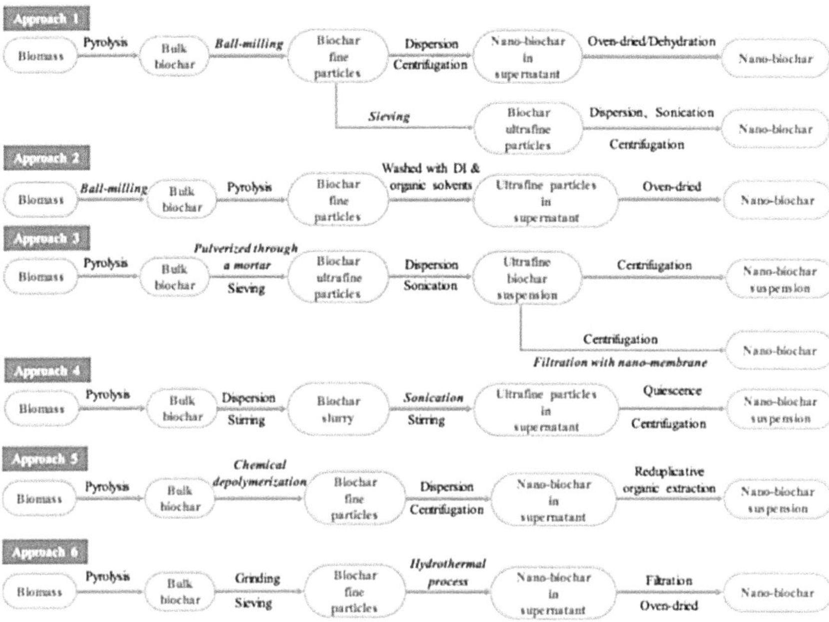

FIGURE 8.4 Commonly used approaches for nano-biochar preparation from biomass.

Source: Reprinted with permission from Ref.[85] Copyright © 2022 Elsevier Ltd.

soil porosity. Bulk density, which represents the mass of the soil per unit volume, is influenced by the properties of NPs. Nanoparticles in soil have been shown to decrease bulk density, increase microporosity, and cause flocculation [43]. Soil particle flocculation reduces macroporosity as well, leading to both tiny and major soil cracks. A key contributor to soil deterioration and crop loss, is field compaction, this effect may be used in agriculture to lessen it.

8.4.4 EFFECT OF NANOMATERIALS ON SOIL WATER HOLDING CAPACITY AND WATER INFILTRATION

Soil water holding capacity refers to how much water a specific soil can store for crop usage. Infiltration, on the contrary, is a measure of the soil's ability to allow water to enter and move through the soil profile. NBC is a type of biochar with lower particle sizes, but its impact on the ability of soil to retain water should be different from that of biochar with bigger particle sizes. In this study, the application of NBC reduced the air entry, which made it simpler for water to leave from soil pores and reduced the soil's ability to hold onto water. Therefore, carbon particle size may be the reason for the difference in soil's ability to store water between NBC and biochar treated soil [45]. As a result, the usage of nano-carbon materials in agriculture is more popular and accepted [46].

When biochar is added to soil, it increases the soil's ability to store water while decreasing the size of macro pores. Adding 1% and 3% of biochar increases soil moisture content when used in fields. The ability to study nanoscale or nanostructured materials as fertilizer transporters or controlled-release vectors for producing so-called agricultural land to retain water, which could increase maize emergence rates, is made possible by nanotechnology. At comparable soil water suction, the moisture in soil containing a significantly large amount of NB content was significantly higher than that in soil with lesser or no NB. A promising trend related to the impact of NB on soil water retention capability has been observed previously. This is because high NB application in soil reduces the large voids but increases small or micro and more pore spaces, which results in reducing the amount of macro pores. In short, NB can help a sandy textured soil retain moisture better by promoting plant root growth and eventually high-water absorption [47].

8.4.5 SOIL SURFACE PROPERTIES

Nano additives are applied to soil directly or as a coating over stabilizing materials like cement. Cement is a crucial stabilizing material that is frequently employed in a variety of construction applications. Even if additives are used, cement functions as a binder and a source of strength unless it is nanoscale. The impact of adding three nanomaterials such as nano MgO, CuO, and clay on the geotechnical characteristics of soft soil was examined [48]. The shear strength, compaction parameters, linear shrinkage, and liquid limit were identified. Each of the nanomaterials' individual additions reduced the soil plastic limit, liquid limit, linear shrinkage, and plasticity index. With an increase in the percentage of nanomaterials, the dry density increased. On the other hand, when nanomaterial doses increased, the ideal moisture contents of the soil-nanomaterial mixes declined. Coating nanoparticles with a variety of materials dramatically modifies their properties and behavior in the environment [49]. Therefore, it is essential to investigate the time-dependent stability of coated NPs. Recently, coated nanoparticles have been employed as adsorbents for a variety of environmental contaminants. Derjaguin, Landau, Verwey, and Overbeek (DLVO) theory well explains the stability of NPs [50]. According to this idea, particle-to-particle repulsion and attraction mechanisms govern the stability of NPs in a variety of media.

These nanoparticles (NPs) have a large specific surface area relative to their volume, and a large percentage of their atoms are located on the surfaces, which increases their reactivity. The van der Waals attractive force, greater strength, and increased heat resistance of the nanoparticles all contribute to their strong aggregate-forming abilities [51]. Nanomaterials can be used in soil conservation because of these characteristics. The primary forces behind biological and geological processes, soil microorganisms are crucial to the ecological functions of soil [52]. In comparison to BC, NB improves the cation exchange capacity, permeability, pore volume, and chemical bonding (Figure 8.2). The impregnation may be accomplished by pre-treating the biomass feedstock with metallic ions, supplementing BC with NB, or creating nanocomposites with metallic nanoparticles, which give them a strong affinity to different contaminants. The properties of NB, such as sorption, physiological efficacy, carbonyl group productivity, CEC, permeability, surface energy, and interface-to-volume ratio, are all higher than simple BC.

8.4.6 SOIL pH

An important element that directly affects soil fertility and health is pH. It is a sign of the soil's acidity or alkalinity. The availability of nutrients for plants depends on the pH of the soil. It has been demonstrated that the type of soil acidity or basicity relies on the soil composition. The ideal pH for plants often ranges between 5 and 7. It has been noted that the buildup of various types of nanoparticles, such as Zn, Ag, Au, and Cu, among others, may have an impact on the soil's pH. It was discovered that the soil's pH also affects how toxic nanoparticles are to worms and bacteria. The impact of organic matter and soil pH on *Folsomia candida* toxicity caused by ZnO nanoparticles. Greater toxicity was discovered to be caused by the soil's pH, which may also affect how insects absorb silver nanoparticles that have accumulated in the soil. It is an important factor that directly affects the fertility and health of the soil. It is an indicator of the soil's acidity or alkalinity. The pH of the soil influences the availability of nutrients for plants. It has been demonstrated that the acidity or basicity of the soil depends on its composition. The optimal pH range for plants is often between 5 and 7. The accumulation of various nanoparticles, including Zn, Ag, Au, Cu, etc., has been reported to affect the soil's pH. The soil's pH was discovered to impact the toxicity of nanoparticles to microorganisms and nematodes.

TABLE 8.1 Effect of Nanoparticles on Soil Microbial Community and Enzymes

Nanomaterials	Major Effect
Au	Enzymatic activities of soil.
TiO_2	Diversity of the bacterial community.
Cu	C, N biomass.
A mixture of Cu and Fe	Microbial community.
A mixture of Co, Fe, Ni, and Ag	Not much effect on the soil community.
Ag	Glucosidase, acid phosphatase, dehydrogenase, urease.
C60	Eucine-aminopeptidase, β-cellobiohydrolase, acid phosphatase, -glucosidase, chitinase, xylosidase.

Source: Reprinted with permission from Ref.[85]

The influence of soil organic matter and pH on the toxicity of ZnO nanoparticles against *Folsomia candida* [53]. It was discovered very recently that copper oxide nanoparticles may be able to alter the pH of paddy soil. Copper oxide nanoparticles enhance the soil's pH, which eventually influences the soil's properties. Insect uptake of silver nanoparticles deposited in soil may also be impacted by soil pH. The pH of the soil appears to be the primary factor governing the amounts of soluble metals. Zn solubility is inversely correlated with pH [54]. In anaerobic digestion, the pH has an impact on NBC and that of the accompanying microorganisms. NBC has an alkaline pH due to the abundance of thermal decomposition of acidic organic compounds. The temperature of pyrolysis and pH value of NBC regularly increase. The NBC also has antioxidant capacity, which allows it to absorb or contribute ions. Furthermore, the existence of heterocyclic and aliphatic groups also facilitates the transport of electrons. NBC enhances anaerobic digestion alkalinity (minimum pH 6), increasing microflora for rapid ammonia production and adaptation to the first loaded stress.

8.4.7 CATION EXCHANGE CAPACITY (CEC)

Clay often has a greater specific surface area than sand and silt, which causes more activity and, as a result, causes clay to absorb more water. In certainty, CEC decreases as the fraction of microsilica in soil that contains lime increases. As a result of a decline in CEC and the formation of bigger particles, water absorption, and swelling will also be reduced. Additionally, as additive content grew, the ability to distinguish between counter-ions (ions with a charge opposed to the fixed charge of the membrane) also improved. TiO_2 nanoparticle loading's impact on the properties of heterogeneous IEM was studied by [55]. In contrast to the monovalent ion (Na_+), they discovered that TiO_2 showed a strong adsorptive affinity toward the bivalent (Ba_{2+}) ion attraction. Compared to sodium, the significant adsorption decreased the permeability of barium.

The key factor controlling the availability of vital nutrients is soil cation exchange capacity (CEC), which NMs may change by interacting with soil texture and soil organic matter. In comparison to the control, the biochar applications reduced the water-soluble electrolyte concentration in the soil, while the potassium content rose in the top 12 to 20 cm of soil and reduced in the subsurface (20 cm) soil. Calcium and magnesium levels rose across the soil profile overall due to modified NB. NB may release the mineral nutrients into the soil. These nutrients are then deposited on clay particles, and as a result, soil with applied NB may have greater nutrient contents. Additionally, the astrocytes (sodium) were quickly replaced by the calcium and magnesium ions, which were released into the soil, lowering the number of soluble electrolytes. Different biochar variants had an impact on how water-soluble salts were distributed.

8.4.8 SOIL HUMIC SUBSTANCES/ORGANIC MATTER/ORGANIC CARBON

Organic matter is the part of the soil that is crucial for the survival of fauna and flora. It is composed of decaying plant and animal matter in various stages, soil microbial cells and tissues, and soil microbe-produced substances. Using nanotechnology to process plant biomass is a novel approach that may help to overcome problems caused by chemical and biological treatments. Nanomaterials have been utilized to improve the synthesis of biochemicals like proteins, lipids, lignin, and carbohydrates. For the use of plant biomass, a variety of nanomaterials including nano photocatalysts, metal oxides, and mesoporous materials have been created. The discovered catalysts are said to provide a method suitable for environmental and economic considerations. However, the necessary procedures have not been created and must be optimized to produce a technique that can be used commercially. Soil health has been defined as the capacity of soil to function as a living framework, within the constraints of the environment and land use, to sustain living system efficiency, maintain or improve water and air quality, and promote plant and living creature health. Soils support a diverse network of soil organisms that contribute to control and safety, mechanical, strategic, and administrative factors. In general, the behavior of nanoparticles has been demonstrated to be highly reliant on environmental factors, such as dissolved organic matter, pH, and ionic strength. For instance, it has been established that soil organic matter quickly adsorbs nanoparticles, considerably enhancing their suspension stability in aqueous solution and their mobility in porous media [56, 57].

Nanomaterials (NMs) may dissolve more quickly if natural organic matter (NOM) works as a chelating agent to bond with the metal ions. For instance, numerous studies have noted that the strong binding affinity between amino acids and Cu ions could cause amino acids to hasten the dissolution of Cu-based NMs. Nano fertilizer can have an impact on soil organisms in both beneficial and harmful ways. They have been discovered to have an impact on both the activities of the soil's microbial community and its structure (Table 8.1). The ecological dangers that nanoparticle exposure poses to soil microorganisms. Although there is no question that nanoparticles can be hazardous to a number of soil-dwelling living forms, some research indicates that nanoparticles may be increasing microbial activity. The biggest carbon storage is SOC, which is two times as abundant as atmospheric carbon. One of the key elements influencing the amount of carbon dioxide in the atmosphere is the conservation or emissions of organic carbon. Although carbon absorption in soil is a lengthy procedure, it is a sustainable natural method of reducing atmospheric CO_2 concentration. According to estimates, even slight variations in SOC may have a big impact on the levels of CO_2 in the atmosphere; for example, a 10% shift in SOC can affect more than 32% of CO_2 concentration. Owing to inappropriate cultivation, agricultural soils often lose a large quantity of SOC. In order to enhance the amount of SOC, different plowing techniques, an appropriate fallow, the use of profound crops, and carbon capture and storage in the soil may all be employed.

Due to their usage, biochar and nano-biochar have these features and will improve the physicochemical characteristics of the soil, which will raise the production rate and production while also reducing SOM and raising and lowering CO_2 from the soil. Because of the stability of its carbon (which has an aromatic structure), these materials, which contain several porous structures, are not biodegradable. As a consequence, microbes are unable to use the carbon inside them as a source of energy. Minerals, macronutrients, and micronutrients are also included in biochar ash. These substances are essential to the chemical and aromatic composition of biochar. For instance, the element nitrogen may be bonded to one or two carbon atoms, although it is unclear how this link would affect how biochar behaves in the soil. Iron-rich biochar is created by heating plant nutrients to temperatures exceeding 600°C, which results in the formation of Fe_3C bonds and ferromagnetic iron clusters. Biochar is created at varying temperatures than the feedstock, and as the heat rises, more persistent carbohydrates are created. The specific surface area of the biochar increases when the temperature is raised by eliminating extraneous particles from the feedstock.

Due to its usage, biochar has qualities that improve the physicochemical characteristics of the soil, which could also increase the growth rate and yield of crops while decreasing soil organic matter and lowering CO_2 levels in the soil. The demand for agricultural goods worldwide is growing. Increased farming activities have led to lower soil quality and increased soil runoff. Some conventional farming practices also result in a reduction in soil organic matter (SOM), which weakens the soil's structural integrity. Consequently, the soil retains fewer minerals and less moisture. When comparing organic farming techniques to traditional ones, these improvements range from 5 to 34% for soil fertility indicators such as SOM and pH.

According to one study, humic substances form strong bonds with nanomaterials that alter the synthetic components of humic materials [37]. SOM is essential for soil structure and carbon cycling as it contains bio-activated soil organic carbon. The rate of carbon breakdown and biological growth in soils is influenced by soil structure. Nanomaterials can potentially alter soil qualities by modifying SOM characteristics. Despite the possibility that certain NMs won't have an impact on SOM, their relationships with other factors might influence how SOM availability and interaction with other processes due to SOM's deposition on NM's outer surfaces. According to the findings, NB is expected to promote SOC sustainability and decrease its decomposition in the soil. The three techniques listed below may improve the sustainability of SOC.

i. Structural preservation of OC by agglomeration;
ii. Integration of OC with soil primary particles especially clay and silt; and
iii. Soil enzymatic stabilization through the development of biodegradable organic carbon compounds.

The addition of NB to soil subsequently results in higher CO_2 gas emissions. In the first twenty days after applying 8% biochar NB to the soil, for instance, CO_2 emissions increased; but, after 4-month time period, it was significantly reduced. Shortly, by ramping up NB, CO_2 is decreased. The physical and metabolic characteristics of NB, soil characteristics, and soil microbes all have a role in the release of CO_2. There is evidence in the literature that natural factors can alter the effects of nanomaterials, particularly those that emit particles. AgNPs are less adaptable in soils with a high natural matter content [58], because natural materials, such as sewage sludge, bind to AgNPs with sorption rates of more than 90% for uncoated particles and lower rates for functionalized particles. Concentrates with biofilm networks have also demonstrated that NPs and natural matter, such as exopolysaccharides and humic acids, can protect microorganisms [59].

8.4.9 AVAILABILITY OF PLANT NUTRIENTS

The mixture and quantity of mineral nutrients present in the soil play a significant role in determining plant growth and development. Fertilizers are a novel facility that can improve nutrient usage effectiveness and lower the cost of environmental pollution. A product that supplies nutrients to crops in the nanoscale range is referred to as a "nano-fertilizer." As an illustration, encapsulation within nanomaterials is covered with a thin protective polymer coating or in the form of nanoscale-sized particles or emulsions. Nanomaterial surface coatings on fertilizer particles hold the material more firmly than traditional surfaces due to higher surface tension, aiding in regulated release. The use of a variety of nanomaterials for beneficial objectives in agricultural production, notably as fertilizers to boost plant output, is becoming increasingly popular. Nanomaterials that are made of the nutrient of interest, like hydroxyapatite nanoparticles for P, can be used as fertilizers.

Agrochemical ingredient delivery, such as fertilizer delivery of macro and micronutrients, is an essential component of using nanotechnology in agriculture. Nitrogen consumption in a rice paddy field with saline-alkali soil was considerably changed by the application of varied doses of nitrogen fertilizer and nanocarbon, and yields rose when nitrogen was added with nanocarbon. Two slow-release fertilizers containing nanocarbon resulted in increased rice yields and improved efficiency of nitrogen utilization. Nitrogen losses in surface water runoff from the paddy soil for both slow-release fertilizers were reduced compared to the fertilizers alone. The release of nitrogen from fertilizers and its uptake by crops might be synchronized by means of nanotechnology. Nano fertilizers should release nutrients on demand while preventing their premature conversion to chemical or gaseous forms that cannot be absorbed by plants.

However, according to a study, biochar can be converted into nano-biochar, which is widely used in soil, flora, and atmospheric cycling processes with appropriate processing. Nano-biochar could closely integrate with mineral components and soil macronutrients to create an enhanced fertilizer due to its large surface area and smaller fragment size. Nano-biochar might prevent the loss of nutrients from water erosion in uneven topography with varying plant cover [60]. The study also concluded that nano-biochar had a substantial impact on soil moisture flow on the Loess Plateau. NBC has the capability to boost soil quality by offering macronutrients, including K_+, more readily available in the rhizosphere [61]. When the absorption is not adequate, K_+ is highly accessible in the soil. The horizontal transport of K_+ was inhibited by the agglomeration of NB on soil fragments, which enhanced the quantity of microscopic pore spaces. By doing so, significant K_+ concentrations are guaranteed in the rhizosphere soil amended with NB, the retention duration of the accessible K_+ increased, and eventually, this enhances the availability and use of K_+ by corn crops. Applying NB to the topsoil seems to have a significant impact on corn development.

The biggest nutritional demands in maize crops are normally expected during the shoot development stage by subsurface plant parts. Compared to the tasseling, grain filling, and ripening phases, mostly high K_+ uptake and retention are observed. Primary nutrient deposition in corn crops progressively declined just after the sprouting phase. Except for nitrogen, NBC is attached to the exchange sites of primary macronutrients. This encourages maize to absorb N, and N is subsequently transferred from the rhizosphere to above-ground plant parts, increasing the N contents in the plants. Consequently, the N uptake of maize plants was higher in the soil amended with NB. Improving the nutrient uptake in the plant in the soil is yet another possible use for NB as a soil supplement, this use is widely documented by NB in recent studies.

NB reduced the seed development of such vegetable crops at a low concentration (50–100 mg/L). In order to reveal the effect of NB on the growth and development of different plants, Chen et al. [60] examined the impact of NB on crop development by using up to 10% NB in the soil in a rain simulation study and found that NB was prevented. In a subsequent study [62], Phosphorus (P) was used as the prototype chemical to examine the uptake capacity of NB in the soil. According to studies, NB enhances P uptake by up to 20%, and this rise is correlated with soil electrolytes. They suggested these results based on the finding of 0.5–1% NBC application. The effects of NB in soil on the production and development of maize were studied by [37]. They discovered that the addition of NB improved the primary nutrient content in the corn in the surface and subsurface soil layers. Nevertheless, the negative impacts of utilizing NB as an application have just been investigated, and they will be covered in the sections that follow. Nutrient concerns include soil degradation associated with subpar agricultural practices, urbanization, desertification, and industrialization; these issues are worldwide growing. It has been shown previously that rising soil disturbances lead to nutritional deficiencies in plants. Lacking essential micronutrients and macronutrients and the resulting loss in production might have a significant impact on global food sustainability and socioeconomic growth. Mineral fertilizers are perhaps the primary popular strategy for enhancing soil's nutritional components.

8.4.10 EFFECT OF NANOMATERIALS ON SOIL REMEDIATION

Due to their high reactivity, high surface-to-volume ratio, surface functionalization, and ability to alter physical parameters like size, shape, porosity, and chemical composition, nanomaterials have recently gained popularity for soil remediation. The combination of these attributes enables efficiency and selectivity in the capture of contaminants. Due to the *in-situ* application, the intercalation of nanoparticles in the soil enables the cleaning of large areas while reducing costs and time. Nanoscale zero-valent iron, carbon nanotubes, and magnetic nanoparticles have dominated soil pollution nano-remediation [63]. Growing research suggests that concern for nanomaterials' manufacture and frequency for the insertion of different modifications is needed to make up for their limitations. Such procedures are an important source of innovation and have been shown to significantly improve the ability of nanoparticles to clean up. Additionally, the development of remediation solutions combining pertinent biological agents and nanomaterials has also shown promise. However, experiments without contaminants have been conducted in order to gain a deeper understanding of the methods for removing impurities and the behavior of nanomaterials in nature. The removal of heavy metals from soil has significantly advanced thanks to nanomaterials, which have also demonstrated effective removal capabilities.

It offers features including a big specific surface area, a lot of adsorption power, and a lot of reactivity. It has enormous promise for removing heavy metals from soil. The majority of recent research on the application of nanomaterials to the removal of heavy metals from soil focuses on the efficiency of the removal process as well as the effects on the soil environment. There is now a study gap on how nanoparticles work in the mending process, but this will change in the future. Heavy metal ions, such as lead, cadmium, mercury, and arsenic, pose the greatest health risks to humans [64]. Various forms of adsorbents formed from carbon-based nanomaterials, metal and magnetic nanocomposites, agricultural and industrial waste, bio adsorbents, and various mineral materials were reported to be employed to remove various heavy metal ions from soils.

8.4.11 NANOMATERIALS FOR PESTICIDES REMOVAL

Nanotechnology has aided in the development of disease-free crops in the last 10 years. Nano silica, a special kind of nanomaterial, is employed as a nano pesticide and falls under the broad category of nanotechnology. Numerous research supports the usage of Si-NPs as a nano pesticide. Si-NPs have been seen to be employed in two different ways: either directly as pesticides in the field, killing insects and larvae, or as mesoporous nanoparticle nanocarriers that release commercial pesticides to increase their effectiveness. Nanoencapsulation involves sealing tiny amounts of the active substances inside a sac or shell with a thin wall (protective coating). Insecticides, fungicides, and nematicides can be nano-encapsulated to create formulations that effectively manage pests while limiting the residual buildup in the soil. A nanotechnology approach of controlled release of the active ingredient may be used to improve the efficacy of the formulation and protect the active ingredient from deterioration and exposure to nanotechnology's uses, which may significantly reduce the amount of pesticide input and associated environmental risks.

Herbicide persistence is increased by effective nano-encapsulated herbicide delivery in the crop-weed environment. Silica is used to create nano-silica, a special kind of nanomaterial. The fact that it can be employed as a nano-pesticide is the most significant of its many uses as a catalyst in medicine and the development of new drugs. The application of nano-silica as a nano-pesticide. The usage of various cuticular lipids by insect pests to protect their water barrier and avoid death from desiccation forms the basis of the process for controlling insect pests using nano-silica. Insects and other pests are another major concern during the cultivation of food crops. As a result, pesticides are commonly used to control pests and diseases transmitted by insects. Certain pesticides are carcinogenic, mutagenic, and teratogenic [65]. It has been demonstrated that pesticides containing organophosphorus compounds reduce human fertility by lowering testosterone levels and producing behavioral issues in offspring and immunological problems [65]. Therefore, the elimination of pesticides is a major priority. Several adsorbents produced from microplastics, bio-nanocomposites, polymer-based composites, zeolites, and MWCNTs have been employed to remove various pesticides from wastewater.

8.5 ALLELOPATHIC EFFECT OF NANOMATERIALS IN AGRICULTURE

The traditional methods for applying agrochemicals to crops include broadcasting and spraying. Nano-encapsulated agrochemicals should be created with all necessary properties, such as effective concentration (with high solubility, stability, and efficacy), time-controlled release in response to certain stimuli, and enhanced targeted action, to minimize the repetitive application, reduce eco-toxicity, and provide a safe and simple mode of delivery. The best example is the use of nano-encapsulated herbicides to suppress parasitic weeds, which lowers the phytotoxicity of herbicides on crops. When nano-capsules are properly functionalized, they can better penetrate the cuticle and enable a delayed, regulated release of the active chemicals once they have reached the desired location. Insects' protective wax layer is harmed when properly functionalized lipophilic nano-silica is introduced into the lipids of the cuticular layer.

An extensive range of industrial and agricultural areas have used nanomaterials. It has been discovered that nanocarbon enhances plant development. They can also be utilized as biomarkers or sensors to identify plant pathogens. Nanomaterials employed in agriculture have been shown to have toxic effects on agroecosystem constituents, including plants and soil bacteria. Studies on the systemic impacts of changed nanomaterial reactivity on plants, people, and microorganisms are still in their infancy. Some of the known hazards related to the usage of nanomaterials include cellular DNA damage, metabolic malfunction, heavy metal release from metallic nanoparticles, formation of reactive oxygen species, etc.

Even though biochar has long been employed as a soil additive to fix contaminants in the soil and prevent them from being ingested by plants, the characteristics of NB have recently been explored. The addition of NB acts as an allelopathic conditioner for soil; it enhances plant development. The detrimental effects of heavy metals on the seedling could be mitigated by applying NB at nominal concentrations. NB improved the variety of soil microorganisms while reducing the absorption of heavy metal (cadmium) by up to 90% in the surface and subsurface plant regions.

8.6 FUTURE PERSPECTIVES AND CHALLENGES

As soils are the ultimate sink for all biotic and abiotic terrestrial chemicals, the accumulation of nanoscale metal-based compounds is inescapable [66–70]. Nanobiotechnology has expedited the majority of agricultural sciences and crop research fields [70–74]. It is widely known that using nanomaterials in the scientific and agricultural fields offers several benefits over conventional practices [75]. Nanotechnology-based biofertilizers are extremely beneficial for plant growth and development. Biopesticides are more effective against diseases and aggressive insects than fungicides. With the use of nanobiotechnology, DNA delivery in plant research has become quite simple [76–78]. Very little is known about how discharged nanomaterials affect the physicochemical qualities of soil and the living forms inhabiting terrestrial ecosystems [79–83]. Therefore, conducting an in-depth study on how nanomaterials interact with biological elements has become important to synthesize effective and non-lethal nanomaterials.

8.7 CONCLUSION

NB and NM hold immense opportunities in agricultural practices as compared to their bulk counterparts. So far, several studies have reflected substantial modifications in soil physicochemical and biological characteristics after the soil addition of NB and NM. The extremely small size of NB, offering increased surface area, imparts them with the potential for environmental remediation of inorganic and organic contaminants, improving soil quality, and hence crop performance. The nano-BC-associated nutrients could be transferred to native soil plants and microorganisms, thereby facilitating plant growth, microbial diversity, and soil biological processes. Although nano-BC possesses greater advantages for agricultural applications compared to biochar, the risks to soil organisms and human health rendered by its small size cannot be denied. Therefore, more field studies would be required to recommend nano-BC for agricultural applications. Investigations on the effect of nano-BC on soil annelids, arthropods, and microbes regulating the cycling of carbon, nitrogen, and sulfur should be carried out extensively for safe application.

KEYWORDS

- **cation exchange capacity**
- **Derjaguin, landau, Verwey, and Overbeek**
- **intrinsic properties**
- **macroaggregate**
- **nano-biochar**
- **nanoparticles**
- **natural organic matter**
- **supermagnetic iron oxide nanoparticles**

REFERENCES

1. Dong, H., Gao, Y., Sinko, P. J., Wu, Z., Xu, J., & Jia, L. (2016). The nanotechnology race between China and the United States. *Nano Today, 11, 7–12.* https://doi.org/10.1016/j.nantod.2016.02.001.
2. Torney, F., Trewyn, B. G., Lin, V. S. Y., & Wang, K. (2007). Mesoporous silica nanoparticles deliver DNA and chemicals into plants. *Nature Nanotechnology, 2,* 295–300. https://doi.org/10.1038/nnano.2007.108.
3. Gao, K., & Jiang, X. (2006). Influence of particle size on the transport of methotrexate across the blood-brain barrier by polysorbate 80-coated polybutylcyanoacrylate nanoparticles. *International Journal of Pharmaceutics, 310,* 213–219. https://doi.org/10.1016/j.ijpharm.2005.11.040.
4. Chausali, N., Saxena, J., & Prasad, R. (2021). Nano biochar and biochar-based nanocomposites: Advances and applications. *Journal of Agricultural and Food Research, 5, 100191.* https://doi.org/10.1016/j.jafr.2021.100191.
5. Chen, X., Duan, M., Zhou, B., & Cui, L. (2022). Effects of biochar nanoparticles as a soil amendment on the structure and hydraulic characteristics of sandy loam soil. *Soil Use and Management, 38, 836–849.* https://doi.org/10.1111/sum.12740.
6. Navarro-Espinoza, S., Meza-Figueroa, D., Guzmán, R., Duarte-Moller, A., Esparza-Ponce, H., Paz-Moreno, F., & Soto-Puebla, D. (2021). Release of nanoparticles in the environment and catalytic converters aging. *Nanomaterials, 11, 3406.* https://doi.org/10.3390/nano11123406.
7. Klaine, S. J., Alvarez, P. J. J., Batley, G. E., Fernandes, T. F., Handy, R. D., Delina, & Lead, J. R. (2008). Critical review nanomaterials in the environment: Behavior, fate, bioavailability, and effects. *Environmental Toxicology and Chemistry, 27.* https://doi.org/10.1002/etc.4147.
8. Wang, H., Zhao, R., Li, Y., Liu, H., Li, F., Zhao, Y., & Nie, G. (2016). Aspect ratios of gold nanoshell capsules mediated melanoma ablation by synergistic photothermal therapy and chemotherapy. *Nanomedicine, 12,* 439–448. https://doi.org/10.1016/j.nano.2015.11.013.
9. Sharma, V. K., Yngard, R. A., & Lin, Y. (2009). Silver nanoparticles: Green synthesis and their antimicrobial activities. *Advances in Colloid and Interface Science, 145,* 83–96. https://doi.org/10.1016/j.cis.2008.09.002.
10. Auffan, M., Rose, J., Proux, O., Borschneck, D., Masion, A., Chaurand, P., & Wiesner, M. R. (2008). Enhanced adsorption of arsenic onto maghemites nanoparticles: As (III) as a probe of the surface structure and heterogeneity. *Langmuir, 24,* 3215–3222. https://doi.org/10.1021/la702998x.
11. Goswami, L., Kim, K. H., Deep, A., Das, P., Bhattacharya, S. S., Kumar, S., & Adelodun, A. A. (2017). Engineered nanoparticles: Nature, behavior, and effect on the environment. *Journal of Environmental Management, 196, 297–315.* https://doi.org/10.1016/j.jenvman.2017.01.011.
12. Liu, C., & Zhang, Z. J. (2001). Size-dependent superparamagnetic properties of Mn spinel ferrite nanoparticles synthesized from reverse micelles. *Chemistry of Materials, 13,* 2092–2096. https://doi.org/10.1021/cm0009470.

13. Scheckel, K. G., Luxton, T. P., el Badawy, A. M., Impellitteri, C. A., Tolaymat, T. M. (2010). Synchrotron speciation of silver and zinc oxide nanoparticles aged in a kaolin suspension. *Environmental Science & Technology, 44*, 1307–1312. https://doi.org/10.1021/es9032265.
14. Bhattacharya, S. S., & Das, S. (2020). Nanomaterials in soil environment: A review. *NASS Journal of Agricultural Sciences, 1*. https://doi.org/10.36956/njas.v1i2.11.
15. Rajput, V. D., Minkina, T., Fedorenko, A., Chernikova, N., Hassan, T., Mandzhieva, S., & Burachevskaya, M. (2021). Effects of zinc oxide nanoparticles on physiological and anatomical indices in spring barley tissues. *Nanomaterials, 11, 1722*. https://doi.org/10.3390/nano11071722.
16. Zhang, J. Y., Zhou, H., Gu, J. F., Huang, F., Yang, W. J., Wang, S. L., & Liao, B. H. (2020). Effects of nano-Fe_3O_4-modified biochar on iron plaque formation and Cd accumulation in rice (*Oryza sativa* L.). *Environmental Pollution, 260, 113970*. https://doi.org/10.1016/j.envpol.2020.113970.
17. Inyang, E. P., Inyang, E. P., Akpan, I. O., Ntibi, J. E., & William, E. S. (2021). Masses and thermodynamic properties of a quarkonium system. *Canadian Journal of Physics, 99*, 982–990. https://doi.org/10.1139/cjp-2020-0578.
18. Xiao, J., Hu, R., & Chen, G. (2020). Micro-nano-engineered nitrogenous bone biochar was developed with a ball-milling technique for high-efficiency removal of aquatic Cd (II), Cu (II), and Pb (II). *Journal of Hazardous Materials, 387, 121980*. https://doi.org/10.1016/j.jhazmat.2019.121980.
19. Oleszczuk, P., Ćwikła-Bundyra, W., Bogusz, A., Skwarek, E., & Ok, Y. S. (2016). Characterization of nanoparticles of biochars from different biomass. *Journal of Analytical and Applied Pyrolysis, 121*, 165–172. https://doi.org/10.1016/j.jaap.2016.07.017.
20. Chen, T., Zhou, Z., Han, R., Meng, R., Wang, H., & Lu, W. (2015). Adsorption of cadmium by biochar derived from municipal sewage sludge: Impact factors and adsorption mechanism. *Chemosphere, 134*, 286–293. https://doi.org/10.1016/j.chemosphere.2015.04.052.
21. Lian, F., Yu, W., Zhou, Q., Gu, S., Wang, Z., Xing, B., & Xing, B. (2020). Size matters: Nano-biochar triggers decomposition and transformation inhibition of antibiotic resistance genes in aqueous environments. *Environmental Science & Technology, 54, 8821–8829*. https://doi.org/10.1021/acs.est.0c02227.
22. Khan, H. A., Naqvi, S. R., Mehran, M. T., Khoja, A. H., Khan Niazi, M. B., Juchelková, D., & Atabani, A. (2021). A performance evaluation study of nano-biochar as a potential slow-release nano-fertilizer from wheat straw residue for sustainable agriculture. *Chemosphere, 285, 131382*. https://doi.org/10.1016/j.chemosphere.2021.131382.
23. Ramadan, M. M., & Abd-Elsalam, K. A. (2020). Micro/nano biochar for sustainable plant health: Present status and future prospects. *Carbon Nanomaterials for Agri-Food and Environmental Applications*, 323–357.
24. Sashidhar, P., Kochar, M., Singh, B., Gupta, M., Cahill, D., Adholeya, A., & Dubey, M. (2020). Biochar for delivery of Agri-inputs: Current status and future perspectives. *Science of the Total Environment, 703, 134892*.
25. Kianfar, E., Abed Hussein, B., Mahdi, A. B., Emad Izzat, S., Acwin Dwijendra, N. K., Romero Parra, R. M., & Mustafa, Y. (2022). Production, structural properties nano

biochar and effects nano biochar in soil: A review. *Egyptian Journal of Chemistry, 65*(12), 607–618. https://doi.org/10.21608/ejchem.2022.131162.5772.
26. Ben-Moshe, T., Dror, I., & Berkowitz, B. (2009). Oxidation of organic pollutants in aqueous solutions by nanosized copper oxide catalysts. *Applied Catalysis B: Environmental, 85*, 207–211. https://doi.org/10.1016/j.apcatb.2008.07.020.
27. Simonin, M., & Richaume, A. (2015). Impact of engineered nanoparticles on the activity, abundance, and diversity of soil microbial communities: A review. *Environmental Science and Pollution Research, 22*, 13710–13723. https://doi.org/10.1007/s11356-015-4171-x.
28. Martin Calvarro, L., de Santiago-Martín, A., Quirós Gómez, J., González-Huecas, C., Quintana, J. R., Vázquez, A., & Ramírez Vera, R. (2014). Biological activity in metal-contaminated calcareous agricultural soils: The role of the organic matter composition and the particle size distribution. *Environmental Science and Pollution Research, 21*, 6176–6187. https://doi.org/10.1007/s11356-014-2561-0.
29. Chandralekha, C. (2022). Synthesis and Characterization of Nanoparticles. In: Ramu, A., Chee Onn, C., Sumithra, M. (eds) International Conference on Computing, Communication, Electrical and Biomedical Systems. EAI/Springer Innovations in Communication and Computing. Springer, Cham. https://doi.org/10.1007/978-3-030-86165-0_7.
30. Ghazi, D., Abbas, A., Abdelghany, A., Elsherpiny, M., & ElGhamry, A. (2022). Evaluating nanotechnology in raising the efficiency of some substances used in fertilizing wheat grown on sandy soil. *Egyptian Journal of Soil Science, 62*, 123–135. https://doi.org/10.21608/ejss.2022.147839.1512.
31. Tourinho, P. S., van Gestel, C. A. M., Lofts, S., Svendsen, C., Soares, A. M. V. M., & Loureiro, S. (2012). Metal-based nanoparticles in soil: Fate, behavior, and effects on soil invertebrates. *Environmental Toxicology and Chemistry, 31*, 1679–1692. https://doi.org/10.1002/etc.1882.
32. Phenrat, T., Saleh, N., Sirk, K., Tilton, R. D., & Lowry, G. V. (2007). Aggregation and sedimentation of aqueous nanoscale zerovalent iron dispersions. *Environmental Science & Technology, 41*, 284–290. https://doi.org/10.1021/es061349a.
33. Moerz, S. T., Kraegeloh, A., Chanana, M., & Kraus, T. (2015). The formation mechanism for stable hybrid clusters of proteins and nanoparticles. *ACS Nano, 9*, 6696–6705. https://doi.org/10.1021/acsnano.5b01043.
34. Duan, M., Liu, G., Zhou, B., Chen, X., Wang, Q., Zhu, H., & Li, Z. (2021). Effects of modified biochar on water and salt distribution and water-stable macro-aggregates in saline-alkaline soil. *Journal of Soils and Sediments, 21, 2192–2202.* https://doi.org/10.1007/s11368-021-02913-2.
35. Yang, Y., Zhou, B., Hu, Z., & Lin, H. (2020). The effects of nano-biochar on maize growth in northern Shaanxi Province on the Loess Plateau. *Applied Ecology and Environmental Research, 18(2), 2863–2877.* https://doi.org/10.15666/aeer/1802_28632877.
36. Zhang, Y., Li, P., Liu, X., Xiao, L., Shi, P., & Zhao, B. (2019). Effects of farmland conversion on the stoichiometry of carbon, nitrogen, and phosphorus in soil aggregates on the Loess Plateau of China. *Geoderma, 351*, 188–196. https://doi.org/10.1016/j.geoderma.2019.05.037.

37. Yang, S., Li, J., Lu, Y., Chen, Y., & Wang, X. (2009). Sorption of Ni (II) on GMZ bentonite: Effects of pH, ionic strength, foreign ions, humic acid, and temperature. *Applied Radiation and Isotopes, 67*, 1600–1608. https://doi.org/10.1016/j.apradiso.2009.03.118.
38. Obia, A., Mulder, J., Martinsen, V., Cornelissen, G., & Børresen, T. (2016). In situ effects of biochar on aggregation, water retention and porosity in light-textured tropical soils. *Soil and Tillage Research, 155*, 35–44. https://doi.org/10.1016/j.still.2015.08.002.
39. Guo, F., Bao, L., Wang, H., Larson, S. L., Ballard, J. H., Knotek-Smith, H. M., & Han, F. (2020). A simple method for the synthesis of biochar nanodots using a hydrothermal reactor. *MethodsX, 7, 101022*. https://doi.org/10.1016/j.mex.2020.101022.
40. Rasa, K., Heikkinen, J., Hannula, M., Arstila, K., Kulju, S., & Hyväluoma, J. (2018). How and why does willow biochar increase a clay soil water retention capacity? *Biomass and Bioenergy, 119*, 346–353. https://doi.org/10.1016/j.biombioe.2018.10.004.
41. Lusiba, S., Odhiambo, J., & Ogola, J. (2017). Effect of biochar and phosphorus fertilizer application on soil fertility: Soil physical and chemical properties. *Archives of Agronomy and Soil Science, 63, 477–490*. https://doi.org/10.1080/03650340.2016.1218477.
42. Lian, F., & Xing, B. (2017). Black carbon (biochar) in water/soil environments: Molecular structure, sorption, stability, and potential risk. *Environmental Science & Technology, 51*, 13517–13532. https://doi.org/10.1021/acs.est.7b02839.
43. Taha, M. R., & Taha, O. M. E. (2012). Influence of nano-material on the expansive and shrinkage soil behavior. *Journal of Nanoparticle Research, 14*. https://doi.org/10.1007/s11051-012-1190-0.
44. Nassaj-Bokharaei, S., Motesharezedeh, B., Etesami, H., & Motamedi, E. (2021). Effect of hydrogel composite reinforced with natural char nanoparticles on the improvement of soil biological properties and the growth of water deficit-stressed tomato plant. *Ecotoxicology and Environmental Safety, 223, 112576*. https://doi.org/10.1016/j.ecoenv.2021.112576.
45. Li, L., Zhang, K., Chen, L., Huang, Z., Liu, G., Li, M., & Wen, Y. (2017). Mass preparation of micro/nano-powders of biochar with water-dispersibility and their potential application. *New Journal of Chemistry, 41*. https://doi.org/10.1039/c7nj00742f.
46. Tan, J. M., Arulselvan, P., Fakurazi, S., Ithnin, H., & Hussein, M. Z. (2014). A review on characterizations and biocompatibility of functionalized carbon nanotubes in drug delivery design. *Journal of Nanomaterials, 2014(34), 20*. https://doi.org/10.1155/2014/917024.
47. Fan, J., Chen, X., Xu, Z., Xu, X., Zhao, L., Qiu, H., & Cao, X. (2020). One-pot synthesis of NZVI-embedded biochar for remediation of two mining arsenic-contaminated soils: Arsenic immobilization associated with iron transformation. *Journal of Hazardous Materials, 398, 122901*. https://doi.org/10.1016/j.jhazmat.2020.122901.
48. Majeed, Z. H., Taha, R., & Jawad, I. T. (2014). Stabilization of soft soil using nanomaterials. *Research Journal of Applied Sciences, Engineering and Technology, 8*, 503–509.

49. Christian, P., von der Kammer, F., Baalousha, M., & Hofmann, T. (2008). Nanoparticles: Structure, properties, preparation, and behavior in environmental media. *Ecotoxicology, 17, 438–447.* https://doi.org/10.1007/s10646-008-0210-4.
50. Wang, H., Adeleye, A. S., Huang, Y., Li, F., & Keller, A. A. (2015). Heteroaggregation of nanoparticles with biocolloids and geocolloids. *Advances in Colloid and Interface Science, 226, 65–82.* https://doi.org/10.1016/j.cis.2015.10.008.
51. Hotze, E. M., Phenrat, T., & Lowry, G. V. (2010). Nanoparticle aggregation: Challenges to understanding transport and reactivity in the environment. *Journal of Environmental Quality, 39(6), 1909–1924.* https://doi.org/10.2134/jeq2009.0462.
52. Jacoby, R., Peukert, M., Succurro, A., Koprivova, A., & Kopriva, S. (2017). The role of soil microorganisms in plant mineral nutrition—Current knowledge and future directions. *Frontiers in Plant Science, 8.* https://doi.org/10.3389/fpls.2017.01617.
53. Waalewijn-Kool, P. L., Diez Ortiz, M., van Straalen, N. M., & van Gestel, C. A. M. (2013). Sorption, dissolution, and pH determine the long-term equilibration and toxicity of coated and uncoated ZnO nanoparticles in soil. *Environmental Pollution, 178*, 59–64. https://doi.org/10.1016/j.envpol.2013.03.003.
54. Sauvé, S., Hendershot, W., & Allen, H. E. (2000). Solid-solution partitioning of metals in contaminated soils: Dependence on pH, total metal burden, and organic matter. *Environmental Science & Technology, 34*, 1125–1131. https://doi.org/10.1021/es990688z.
55. Hosseini, Z. S., Shasti, M., Ramezani Sani, S., & Mortezaali, A. (2016). Photodetector diode based on thermally oxidized TiO_2 nanostructures/p-Si heterojunction. *Journal of Applied Physics, 119, 014503.* https://doi.org/10.1063/1.4937546.
56. Xie, B., Xu, Z., Guo, W., & Li, Q. (2008). Impact of natural organic matter on the physicochemical properties of aqueous C_{60} nanoparticles. *Environmental Science & Technology, 42*, 2853–2859. https://doi.org/10.1021/es702231g.
57. Johnson, R. L., Johnson, G. O. B., Nurmi, J. T., & Tratnyek, P. G. (2009). Natural organic matter enhanced the mobility of nano zerovalent iron. *Environmental Science & Technology, 43*, 5455–5460. https://doi.org/10.1021/es900474f.
58. Coutris, C., Joner, E. J., & Oughton, D. H. (2012). Aging and soil organic matter content affect the fate of silver nanoparticles in soil. *Science of the Total Environment, 420*, 327–333. https://doi.org/10.1016/j.scitotenv.2012.01.027.
59. Roh, J. Y., Park, Y. K., Park, K., & Choi, J. (2010). Ecotoxicological investigation of CeO_2 and TiO_2 nanoparticles on the soil nematode *Caenorhabditis elegans* using gene expression, growth, fertility, and survival as endpoints. *Environmental Toxicology and Pharmacology, 29*, 167–172. https://doi.org/10.1016/j.etap.2009.12.003.
60. Chen, X., Zhou, B., Wang, Q., Tao, W., & Lin, H. (2020). Nano-biochar reduced soil erosion and nitrate loss in sloping fields on the Loess Plateau of China. *Catena, 187, 104346.* https://doi.org/10.1016/j.catena.2019.104346.
61. Rajput, V. D., Minkina, T., Ahmed, B., Singh, V. K., Mandzhieva, S., Sushkova, S., Bauer, T., Verma, K. K., Shan, S., van Hullebusch, E. D., et al. (2022). Nano-biochar: A novel solution for sustainable agriculture and environmental remediation. *Environmental Research, 210, 112891.* https://doi.org/10.1016/j.envres.2022.112891.
62. Mahmoud, E., el Baroudy, A., Ali, N., & Sleem, M. (2020). Spectroscopic studies on the phosphorus adsorption in salt-affected soils with or without nano-biochar

additions. *Environmental Research, 184, 109277.* https://doi.org/10.1016/j.envres.2020.109277.

63. Mukhopadhyay, R., Sarkar, B., Khan, E., Alessi, D. S., Biswas, J. K., Manjaiah, K. M., ... & Ok, Y. S. (2022). Nanomaterials for sustainable remediation of chemical contaminants in water and soil. *Critical Reviews in Environmental Science and Technology, 52*(15), 2611–2660.
64. Khan, F. S. A., Mubarak, N. M., Khalid, M., Walvekar, R., Abdullah, E. C., Mazari, S. A., Nizamuddin, S., & Karri, R. R. (2020). Magnetic nanoadsorbents' potential route for heavy metals removal—A review. *Environmental Science and Pollution Research, 27*, 24342–24356. https://doi.org/10.1007/s11356-020-08711-6.
65. Ma, G., & Ms, K. (2016). The effect of pesticide pollution on our life and environment. *Journal of Pollution Effects & Control, 04.* https://doi.org/10.4172/2375-4397.1000159.
66. Rajput, V. D., Minkina, T., Kumari, A., Harish, Singh, V. K., Verma, K. K., Mandzhieva, S., Sushkova, S., Srivastava, S., & Keswani, C. (2021). Coping with the challenges of abiotic stress in plants: New dimensions in the field application of nanoparticles. *Plants, 10*(6), 1221. https://doi.org/10.3390/plants10061221.
67. Rajput, V., Minkina, T., Sushkova, S., Behal, A., Maksimov, A., Blicharska, E., Ghazaryan, K., Movsesyan, H., & Barsova, N. (2019). ZnO and CuO nanoparticles: A threat to soil organisms, plants, and human health. *Environmental Geochemistry and Health, 42*(1), 147–158. https://doi.org/10.1007/s10653-019-00317-3.
68. Rajput, V. D., Minkina, T., Sushkova, S., Tsitsuashvili, V., Mandzhieva, S., Gorovtsov, A., Nevidomskyaya, D., & Gromakova, N. (2017). Effect of nanoparticles on crops and soil microbial communities. *Journal of Soils and Sediments, 18*(6), 2179–2187. https://doi.org/10.1007/s11368-017-1793-2.
69. Rajput, V., Minkina, T., Fedorenko, A., Sushkova, S., Mandzhieva, S., Lysenko, V., Duplii, N., Fedorenko, G., Dvadnenko, K., & Ghazaryan, K. (2018). Toxicity of copper oxide nanoparticles on spring barley (*Hordeum sativum* distichum). *Science of The Total Environment, 645*, 1103–1113. https://doi.org/10.1016/j.scitotenv.2018.07.211.
70. Rajput, V. D., Kumari, A., Minkina, T., Barakhov, A., Singh, S., Mandzhieva, S. S., Sushkova, S., Ranjan, A., Rajput, P., & Garg, M. C. (2022). A practical evaluation of the integrated role of biochar and nanomaterials in soil remediation processes. *Environmental Geochemistry and Health, 45, 9435–9449.* https://doi.org/10.1007/s10653-022-01375-w.
71. Rajput, V. D., Faizan, M., Upadhyay, S. K., Kumari, A., Ranjan, A., Sushkova, S., Chauhan, P. K., Mahmud, A. A., Burachevskaya, M., Chaplygin, V., Deryabkina, I. (2022). Influence of nanoparticles on the plant rhizosphere microbiome. In *Sustainable Plant Nutrition in a Changing World* (pp. 83–102). https://doi.org/10.1007/978-3-030-97389-6_4.
72. Rajput, V. D., Singh, A., Minkina, T., Rawat, S., Mandzhieva, S., Sushkova, S., Shuvaeva, V., Nazarenko, O., Rajput, P., Komariah, Verma, K. K., Singh, A. K., Rao, M., & Upadhyay, S. K. (2021). Nano-enabled products: Challenges and opportunities for sustainable agriculture. *Plants, 10*(12), 2727. https://doi.org/10.3390/plants10122727.
73. Singh, A., Singh, A. K., Rawat, S., Pal, N., Rajput, V. D., Minkina, T., Sharma, R., Singh, N. P., & Tripathi, J. N. (2022). Satellite-based quantification of methane

emissions from wetlands and rice paddies ecosystems in North and Northeast India. *Hydrobiology, 1*(3), 317–330. https://doi.org/10.3390/hydrobiology1030023.

74. Singh, A., Sengar, R. S., Rajput, V. D., Minkina, T., & Singh, R. K. (2022). Zinc oxide nanoparticles improve salt tolerance in rice seedlings by improving physiological and biochemical indices. *Agriculture, 12*(7), 1014. https://doi.org/10.3390/agriculture12071014.
75. Singh, A., Sengar, R. S., Sharma, R., Rajput, P. (2021). Nano-priming technology for sustainable agriculture. *Biogeosystem Technique, 8*(2), 79–92. https://doi.org/10.13187/bgt.2021.2.79.
76. Singh, A., Mehrotra, R., Rajput, V. D., Dmitriev, P., Singh, A. K., Kumar, P., Tomar, R. S., Singh, O., Singh, A. K. (2022). Geoinformatics, artificial intelligence, sensor technology, big data. In *Sustainable Agriculture Systems and Technologies* (pp. 295–313). https://doi.org/10.1002/9781119808565.ch14.
77. Singh, A., Rajput, V. D., Rawat, S., Sharma, R., Singh, A. K., Kumar, P., Singh, A. K., Minkina, T., Singh, R. P., Singh, S. (2022). Geoinformatics and nanotechnological approaches for coping up abiotic and biotic stress in crop plants. In *Sustainable Agriculture Systems and Technologies* (pp. 337–359). https://doi.org/10.1002/9781119808565.ch17.
78. Singh, A., Rajput, V. D., Rawat, S., Sharma, R., Kumar Singh, A., Kumar Singh, A., Tomar, R. S. (2021). Emerging tools for sustainable agriculture and food security: An introduction. In *Emerging Tools for Sustainable Agriculture and Food Security* (Vol. 1, pp. 3–16). Deepika.
79. Rajput, P., Rajput, V. D., Singh, A., Sharma, R., Rawat, S., Tomar, R. S., Singh, A., Minkina, T. (2021). Impact of emerging tools for solving ‘hidden hunger’ in humans. In *Emerging Tools for Sustainable Agriculture and Food Security* (Vol. 1, pp. 17–30). Deepika.
80. Rajput, V. D., Minkina, T., Kumar, A., Shende, S. S., Ranjan, A., Faizan, M., Barakhov, A., Gromovik, A., Gorbunova, N., Rajput, P., Singh, A., Khabirov, I., Nazarenko, O., Sushkova, S., Kizilkaya, R. (2022). A review on nanobioremediation approaches for restoration of contaminated soil. *Eurasian Journal of Soil Science (EJSS), 11*(1), 43–60. https://doi.org/10.18393/ejss.990605.
81. Rajput, V. D., Minkina, T., Feizi, M., Kumari, A., Khan, M., Mandzhieva, S., Sushkova, S., El-Ramady, H., Verma, K. K., Singh, A., van Hullebusch, E. D., Singh, R. K., Jatav, H. S., Choudhary, R. (2021). Effects of silicon and silicon-based nanoparticles on rhizosphere microbiome, plant stress and growth. *Biology, 10*(8), 791. https://doi.org/10.3390/biology10080791.
82. Singh, A., et al. (2020). Monitoring soil salinity and recent advances in the mechanism of salinity tolerance in plants. *Biogeosystem Technique, 7*(2), 66–87. https://doi.org/10.13187/bgt.2020.2.66.
83. Singh, A., Sengar, R. S., Shahi, U. P., Rajput, V. D., Minkina, T., & Ghazaryan, K. A. (2022). Prominent effects of zinc oxide nanoparticles on roots of rice (*Oryza sativa* L.) grown under salinity stress. *Stresses, 3*, 33–46. MDPI AG. https://doi.org/10.3390/stresses3010004.
84. Javed Z., Dashora K., Mishra M., Fasake V.D., and Srivastva A. (2019) Effect of accumulation of nanoparticles in soil health- a concern on future. Front Nanosci Nanotech 5: DOI: 10.15761/FNN.1000182

85. Song B., Cao X., Gao W., Aziz S., Gao S, Lam C, Lin R., 2022. Preparation of nano-biochar from conventional biorefineries for high-value applications, Renewable and Sustainable Energy Reviews, 157, 112057. https://doi.org/10.1016/j.rser.2021.112057.
86. Abd-Elsalam, K. A. (Ed.). (2019). Micro/nano biochar for sustainable plant health: Present status and future prospects Book. Carbon Nanomaterials for Agri-Food and Environmental Applications. Elsevier.

Index

A

Abiotic stresses, 60, 76, 132
Acidobacteria, 73
Actin filaments, 128
Actinobacteria, 73, 78, 186
Actinomycetes, 61, 68
Adsorption, 6–8, 10, 16, 17, 19, 36, 38, 40–42, 47, 61, 69, 72, 97, 100, 101, 104, 117–119, 134, 136, 141, 155, 158, 161, 164, 181, 182, 185–187, 208, 214, 216, 221, 227
Aggregation, 74, 102, 213
Agricultural sector, 2, 118, 119, 129, 132–135, 137, 141, 207
Agriculture application, 167
Algae, 61, 209
Aluminum, 64, 104
  alloys, 64
Amazon
  basin, 179
  delta dark earth, 179
Amphiphilic polyurethane (APU), 75
Anaerobic digestion, 17
Antibacterial
  activity, 132
  qualities, 132
Antioxidant defense system, 131, 132, 134
Antiport transport, 126
Anti-porters, 124
Aquaporins, 126
Arabidopsis K transporter (AKT1), 126
Arid regions, 122
Ash, 99
  content, 99

B

Bacteria, 34, 61, 67, 68, 102, 140, 158, 177, 190, 219, 229
Bacteroidetes, 78, 186
Ball
  milled magnetic nano biochar (BMBCs), 102, 106
  milling, 6–8, 14–16, 18, 19, 36–38, 62–64, 120
Bimetallic nanoparticles, 69
Bioaccumulation, 80, 183
Biocatalytic enzymes, 132
Biochar (BC), 1–11, 14–19, 33–47, 49, 60, 64, 77, 78, 82, 95–106, 117–122, 134–136, 139–141, 153–167, 177–194, 205–219, 221, 223–225, 229, 230
  based nanocomposites (BNC), 39, 40, 135, 194
  particles, 37, 38, 136
  yield, 157
Biofuels, 98
Biological synthesis, 67
Biomass, 2, 4–6, 8, 14, 18, 19, 34, 35, 37, 39–42, 44–46, 61, 77, 96–99, 118, 130, 131, 135, 153, 154, 156, 157, 159, 163, 165, 178–183, 186, 188, 189, 192, 193, 207, 211, 217, 219–222
Biomedical, 68, 69, 76
Biopesticides, 229
Biotic stress, 132, 190
Black carbon, 192
Brassica chinensis L, 44, 78
Brassinosteroids (BR), 127
Brunauer, Emmett, and Teller (BET), 10, 15, 19
Bulk
  biochar (B-BC), 3
  density, 158, 217
Byproduct, 157

C

C mineralization, 130
C potential, 179

Cadmium (Cd), 44, 45, 69, 78, 101, 102, 185, 186, 189, 227, 229
  contaminated soil, 45, 186
Calcineurin B-like (CBL), 126, 141
  interacting protein kinases (CIPKs), 126
Calcium, 124, 221
  carbonate (CaCO3), 15, 122, 193
  permeable channel, 124
Calmodulin (CaM), 125, 141
Canola plants, 131
Carbamazepine, 164
Carbon, 1, 2, 66, 70, 72, 73, 96–98, 100, 153, 192, 193, 208
  based nanoparticles (CBN), 70, 71, 73
  dioxide (CO2), 98
    absorption, 160
    emissions, 96, 177, 224
    levels, 223
  dots, 73
  nanofibers (CNFs), 67, 82
  nanomaterials, 153
  nanoparticles, 72
  nanotube (CNT), 3, 18, 19, 37, 42, 49, 64, 66, 67, 71, 82, 130, 180–183, 194, 208, 227
  rich nano biochar, 98
  sequestration, 96, 100, 193
Carbonization, 157
  process, 157
Carya, 38
Catharanthus roseus, 76
Cation exchange capacity (CEC), 2, 10, 16, 18, 19, 104, 106, 155, 161, 167, 179, 187, 194, 206, 209, 210, 212, 219, 221, 230
C-cycling, 178
Cell wall, 99, 125, 126, 128, 141
Cellulose, 157
  decomposition, 157
Cement, 218
Char, 154, 193
Chemical
  fertilizers, 161, 189
  vapor deposition (CVD), 62, 65, 66, 69, 82
Chitosan, 76
  nanoparticles, 76, 132
Chlorophyll
  biosynthesis, 132
  synthesis, 132
Circadian clock, 126
Clay, 164, 221
  biochar composite, 164
Climate change, 1, 16, 33, 96, 118, 122, 129, 153, 155, 165, 178
Colloidal biochar, 34, 38, 49
Contaminant removal, 19
Copper (Cu), 133, 220
  (II), 101, 164, 186
  ions, 222
  oxide nanoparticles, 220
Cortical microtubules, 128
C-rich adsorbent, 180
Crop
  production, 163
  yield, 123
Current challenges, 33
Cyclic nucleotide-gated ion channels, 125
Cyclin, 129
Cylindrica, 45, 46
Cytosol, 126–128

D

Dephosphorylation, 126
Derjaguin, Landau, Verwey, and Overbeek (DLVO), 219, 230
Development, 3, 17, 33, 34, 36, 37, 59, 60, 67, 69, 73, 76, 103, 119, 123, 124, 129, 131, 132, 134–136, 160, 177–180, 182, 186, 189–191, 194, 209, 211, 213, 214, 224–229
Distilled water, 38, 39, 42
Dry milling, 7
Dynamic soil organic matter (DSOM), 159, 167

E

Electrical conductivity (EC), 10, 16, 104, 106, 117, 120, 123, 141, 161, 167, 187, 194, 206, 209
Environmental application, 17, 166
Enzymatic activities, 220
Enzymes, 220

E-waste recycling facilities, 189
Exogenous application, 132

F

Fabrication, 1, 40, 41, 43
  methods, 40, 41, 43
Fast pyrolysis, 5, 157
Fate and mechanism, 19
Feedstock, 4–7, 9, 10, 14, 16, 18, 19, 42, 68, 96, 97, 105, 118, 120, 156–158, 164, 185, 186, 192, 211, 219, 223
Ferralsols, 178
Fertilizer, 36, 37, 59, 72, 73, 81, 104, 105, 118, 119, 132, 136, 141, 161, 163, 179, 185, 187, 189, 191, 194, 211, 224–226
Ferulic acid (FA), 46
Flash pyrolysis, 5
Folsomia candida, 219, 220
Food packaging, 132
Fourier transform infrared spectroscopy, 10
Free valence electrons, 65
Functional
  groups, 2, 7, 10, 15, 16, 19, 71, 73, 97, 98, 104, 120, 134, 155, 158, 181, 182, 188, 216
  nano-biochar, 183
  nanocomposites, 72
  nanoparticles, 183

G

Gaseous product, 157
Gasification, 5, 6, 156, 157, 165
Gemmatimonadetes, 73
Gene expression, 126, 127, 132, 141
Geologic sequestration, 97
Gliricidia
  sepium, 38
  septum, 120
Global climate change, 153
Glyphosate (GL), 101
Graphene, 71, 164
  oxide, 71
Green revolution, 33
Greenhouse gas emissions, 96

H

Heavy metal (HM), 2, 44, 59, 68, 78, 81, 97, 101, 102, 104, 106, 155, 163, 166, 178, 180, 185, 186, 188, 189, 194, 227, 229
  ions, 227
  polluted land, 44
Hemicellulose, 157
Herbicide persistence, 228
High surface area, 191
Homeobox HD-ZIP Class III (HD-ZIPIII), 128
Human civilization, 33
Hydraulic conductivity, 215
Hydrogen, 125
  peroxide-induced Ca2+ increases 1 (HPCA1), 125
Hydrothermal carbonization, 5, 157
Hydroxide-biochar composites, 39
Hydroxidized biochar, 183
Hyperaccumulators, 61

I

Imperata cylindrica, 45
Industrialization, 160, 187
Infiltration, 217
Insecticides, 132, 184, 194, 228
Interaction, 18, 19, 45, 63, 131, 136, 158, 192, 194, 224
Intrinsic properties, 207, 230
Ionic
  balance, 125, 126
  imbalance, 126, 129, 141
Iron (Fe), 38, 135, 164, 213, 223
  oxide, 38, 135, 213
    nanoparticles, 213
    permeated mesoporous rice-husk nano-BC (IPMN), 38
  rich biochar, 223
Irrigation, 123

J

Jasmonate (JA), 127, 131, 134
  octadecanoid pathway, 134
Jasmonic acid, 131, 134

Judicious selection, 18
Junction, 141

K

Ketone, 15, 98
  groups, 15
Key
  challenges, 8
  contributor, 217
  element, 122, 126, 222
  factor, 221
  features, 130
  functions, 81
  plant, 179
  problems, 80
  role, 6, 130
  soil, 209
Keystone taxa, 209
Killing insects, 227
Kinase, 125–129
  cascade reactions, 127
  SOS2, 125
Kinetic
  constraints, 80
  energy, 6, 213
  restrictions, 74
  studies, 80
Knowledge, 18

L

Leaf water transpiration, 126
Leucine-rich repeat extensins (LRX), 128, 141
Lignin, 157
  decomposition, 157
Lignocellulosic biomass, 157
Liquid product, 157
Lithography, 64
Loess plateau, 225

M

Macroaggregate (MA), 213, 214, 230
Macrospores, 99
Magnesium carbonates, 122, 193
Magnetic
  biochar composites, 183
  nano biochar, 164, 183, 186
Mammalian cells, 18, 19
Manganese (Mn), 134, 164
Materials, 8, 121
Mechanical
  grinding, 101
  milling, 63
Mechanism, 157, 162
Mesoporous, 99
Metal, 186
  ions, 40
  oxide, 69, 80
  precursors, 68
  profiles, 19
Methylene blue, 164
Micelles, 70
Micro/nano biochar, 182
Microbial
  biomass, 159
  community, 220
  population, 103, 130, 132, 158, 184, 191, 192, 194, 210, 212
Micrometer porosity, 99
Micronutrient, 132, 133
Microorganisms, 78
Microporous, 99
Mineral, 15, 17, 19, 34, 45, 60, 81, 98, 100, 103, 117, 122, 134–136, 178, 187–191, 211, 215, 223
  fertilizers, 226
Mitogen-activated protein kinases (MAPK), 127, 141
Modified montmorillonite nano clay, 213
Monometallic nanoparticles (MNPs), 69
Municipal solid waste (MSW), 1, 2, 34, 49, 178
Mutations, 126

N

N intake, 189
N uptake, 226
N2O emission, 161, 180, 187, 194
Na+ Cl– ion concentration, 130
Na+ homeostasis, 126
Na+ ions, 76, 119, 123–125, 132, 136
Na+/H+ antiporters, 125
Na+/K+ homeostasis, 126

NAC domain, 128
NaCl concentration, 131
Nano additives, 218
Nano biochar, 35, 36, 46, 47, 49, 97–102, 105, 106, 153, 163, 164, 166, 179, 182, 214
Nano cerium oxide, 164
Nano fertilizer, 222, 225
Nano silica, 227
Nano-BC (N-BC), 1, 3, 4, 11, 17, 19, 33, 34, 36, 37, 43–45, 47–49, 59, 60, 64, 78, 82, 95–97, 99–101, 105, 106, 117–121, 134, 136, 139, 141, 153, 154, 160, 163–167, 177, 182–184, 186, 192, 194, 205–207, 211–213, 216, 217, 223, 225, 230
  composites, 49, 100, 182
Nanobiotechnology, 229
Nanocomposite, 7, 11, 19, 34, 37, 39, 40, 42, 72, 97, 98, 101, 102, 106, 135, 164, 177, 181–184, 186, 193, 194, 219, 227
Nanoencapsulation, 228
Nanomaterial (NM), 33, 37, 39, 42, 44, 61–67, 70–73, 80, 117–119, 130, 131, 133–138, 140, 141, 153, 161, 181, 182, 184–186, 188, 194, 206–213, 215–229, 230
Nano-metal oxide, 40
  biochar composites (NMOBCs), 40, 49
Nanometer scale, 33
Nanoparticle (NP), 1, 8, 9, 33, 34, 37, 39–44, 59–76, 79–82, 95, 117, 118, 131–135, 137, 140, 141, 153, 161, 164–166, 177, 180–183, 189, 205–209, 211–213, 216–222, 225, 227, 229, 230
  concentration, 80
Nanoscale, 227
  zerovalent iron, 227
Nanosphere lithography (NSL), 64, 82
Nanostructured activated biochar (nPPAB), 47
Nanotechnological approaches, 60, 77
Nanotechnology, 61, 130, 141, 205, 206, 227, 229
  based biofertilizers, 229
Natural organic matter (NOM), 101, 215, 222, 230
Negative modulators, 126
Nitrogen, 72, 225
  consumption, 225
  loss, 225
Nitrous oxides (NOx), 154

O

Organic
  Carbon (OC), 35, 49, 96, 102, 104, 106, 154, 159, 180, 188, 206, 210–212, 222–224
  material, 46
  matter, 221, 159
Oryza sativa, 126
Osmolyte
  accumulation, 132
  biosynthesis, 126
Osmotic
  homeostasis, 125
  potential, 104, 127, 129, 141
  signaling pathway, 126
  stress, 117, 119, 124–126
Oxytetracycline (OTC), 46, 47, 101, 106

P

pH, 3, 15, 17, 18, 36, 37, 67, 68, 75, 77, 97, 104, 118, 135, 154, 161, 187, 188, 191, 206, 209, 210, 212, 216, 219, 220, 222, 223
Pharmaceutical, 46, 69, 101, 106, 132, 178
Phenanthrene (PHEN), 76, 181
Phosphate anions (H2PO4), 161
Phosphorus (P), 226
Phosphorylation, 126
Photodynamic therapy (PDT), 71, 82
Photosynthetic bacteria (PSB), 102
Physicochemical properties, 1, 3, 33, 77, 95, 106, 129, 130, 136, 161, 209
Physiological performance, 132
Phyto nanotechnology, 68
Pinewood, 164
Plant, 17, 36, 37, 44, 45, 60, 61, 68, 72, 76, 77, 96, 97, 99, 104–106, 117–119, 123–136, 138–141, 154, 161–163, 180, 186, 187, 190, 191, 193, 194, 206, 207, 209, 210, 219, 220, 225, 226, 229, 230

growth, 19, 33, 34, 44, 73, 80, 95, 105, 117, 123, 124, 127–129, 131, 133–136, 140, 153, 158, 160–162, 179, 186, 187, 190, 192, 194, 206, 210, 224, 229, 230
metabolism, 131
systems, 182
Plantlet regeneration, 132
Plasma membrane, 125, 126, 128
Platinum, 69
Pollutants, 75, 164
Pollution, 205
Poly(ethylene glycol)-modified urethane acrylate (PMUA), 75
Polychlorinated biphenyls (PCBs), 37, 49, 97
Polycyclic aromatic hydrocarbons (PAHs), 6, 19, 37, 49, 75, 76, 81, 97, 98
desorption, 75
profiles, 105
Polymer nanoparticles, 75
Polymetallic nanoparticles, 69
Polyols, 126
Polyurethane acrylate anionomer (UAA), 75
Population
explosion, 95
growth, 178
Post-pyrolysis, 183
Pottasium (K+)
absorption, 127
deficiency, 126
ions, 126, 129
Precursor type, 67
Pre-treatment, 183
Proline, 126, 129, 132
synthesis, 132
Protein, 68, 69, 80, 125, 126, 134, 141, 222
functions, 132
phosphatase 2C, 126
Proteobacteria, 73
Pure
alcohol, 39
magnetic biochar (PMBCs), 102
Pyrolysis, 2–6, 10, 14, 15, 18, 34–37, 39–42, 62, 65, 67, 69, 75, 96, 98, 105, 117, 118, 120, 153, 154, 156–158, 164, 167, 177, 179–181, 183, 193, 207, 209, 220

Q

Quality, 33
Quick solidification, 69
Quinoa, 134

R

Raw biochar, 5
Reactive
organic species (ROS), 97, 106, 124, 125, 127, 128, 134
oxygen species, 124, 127, 141, 229
Reduction, 18
References, 157, 163, 164, 186
Rice straw, 164
Risk assessment, 2, 19
RNA synthesis, 129
Root
development, 213
growth, 103
water uptake, 126

S

Safer application, 19
Salinity, 76, 117, 123, 135, 138, 139
stress, 76, 117–119, 124, 126, 127, 131–133, 135, 138, 139, 141
Salinization, 123, 129, 130
Salt
stress, 76, 81, 104, 119, 124–130, 132–136, 140, 153
tolerance, 127, 131–133, 140
Scanning electron microscopy, 10
Seed germination, 119, 129, 131, 132, 163
Shoot development, 225
Silica, 131, 228
Silicon, 131
Silt, 102
Slow pyrolysis, 4, 157
S-nitrosomercaptosuccinic acid, 132
Sodification, 123
Soil, 1, 33, 34, 45, 59, 60, 81, 95, 104, 117, 122, 123, 153, 158, 159, 177, 205, 206, 210, 212, 214, 216, 217, 220, 222, 224
amendment, 216
classification, 123
degradation, 34, 47, 226
enzymatic stabilization, 224

erosion, 45
health, 59, 81, 222
microaggregates, 214
microbe interaction, 122, 194
organic
carbon (SOC), 35, 49, 96, 102–104, 106, 193, 212, 222–224
matter, 159, 206, 210
productivity, 106, 130, 190, 209
qualities, 33, 36, 186, 223
remediation, 47, 49, 61, 62, 158, 160, 188, 226
salinization, 122
systems, 19
water holding capacity, 217
Solanum tuberosum, 136
Sol-gel technique, 67, 69
Solid product, 157
Soluble salts, 129
Somatic embryogenesis (SE), 132
Sonication, 8
Specific surface area (SSA), 2, 10, 14, 15, 18, 19, 34, 36, 40, 49, 77, 97, 99, 122, 140, 180, 183, 184, 194, 216, 219, 221, 223, 227
Spinning, 62
Spray pyrolysis, 69
Stability, 159
Stomatal closure, 125, 126
Streptomyces, 18
Structural
preservation, 224
size, 33
Sucrose nonfermenting-1-related protein kinase 2s (SnRK2s), 127
Sulfur dioxide, 154
Supermagnetic iron oxide nanoparticles (SPION), 208, 230
Surface, 74, 158, 191
area, 191
properties, 74
Sustainability, 33
Sustainable, 1, 2, 9, 17, 19, 46, 47, 49, 60, 61, 77, 81, 117–120, 133, 153, 154, 166, 167, 177, 179, 189, 193, 211, 222
choices, 19
technologies, 19
Syngas, 5, 98, 156
Synthesis, 38, 194

T

Technology, 33, 59, 177, 205
Temperature, 6, 157
Tensile strength, 103
Terra-preta, 155
Tetracycline (TC), 102, 164
Thermal decomposition, 62, 63, 98
Titanium dioxide nanoparticles, 80
Tomato, 163
Torrefaction, 156
Transporters, 126, 180, 218
Turgor pressure, 126, 128

U

Ultimate sink, 229
Ultrasonic vibrating, 36
Ultrasonicator, 42
Ultrasound irradiation, 8
Unbalanced chemical fertilizer, 46
Uncoated particles, 224
Unconfined compressive strength, 213, 215, 216
Underwent aqueous, 41
Undoubtedly, 191
Uneven topography, 225
Unfit, 60, 75, 76, 78
Uniform, 7, 64
dispersion, 73
molecular structure, 73
surface, 8
Unintended addition, 178
Unique, 64, 69–71, 130, 131, 153
features, 69
properties, 70, 140, 153
Unmanageable soil, 130
Unproductive, 122, 130
Unshared pair, 76
Unstable, 80, 103
Untreated
biochar, 181
plots, 44, 45
soils, 44, 78
Upholding social equality, 61
Upregulating genes, 76
Upstream, 128

Uptake, 36, 44, 45, 81, 117, 119, 126, 129, 135, 140, 161, 187, 190, 210, 220, 225, 226
Urban areas, 178
Urbanization, 178, 226
Urea, 39, 72, 189
Urease, 220
Urethane, 75
Urgent need, 2
Urine, 46, 47, 178
Utility, 184
Utilization, 36, 166, 209, 211, 225
Utilize, 76, 104
Utilized charcoal, 179
Utilizing biochar, 178, 188

V

Van Krevelen diagrams, 15
Vesicles, 70
Vigna radiata, 73, 134
Volatile organic compounds (VOCs), 15, 190, 194
Volatilization, 81

W

Warming, 41, 122, 193
Waste, 6, 35, 37, 46, 67, 96, 178, 183–185
  biochar, 209
  biomass, 2, 18, 178
  generation, 7, 178
  management, 37, 100
  material, 1, 38, 97, 101
  products, 35, 180
  pyrolysis temperature, 180
Wastewater, 16, 17, 47, 95, 100–102, 155, 165, 182, 228
  bioremediation, 102
  treatment, 17, 35, 95, 100, 101, 165
Water, 119, 123, 193
  absorption, 73, 131, 218, 221
  availability, 158
  bodies, 78, 161, 185
  cleansing, 61
  contamination, 102
  content, 6, 215, 216
  crisis, 60
  deficiency, 119, 126, 128
  erosion, 159, 225
  flowing downstream, 123
  holding
    capacity, 34, 117, 136, 161, 179, 187
    characteristics, 10
    retention capacity, 45
  infiltration, 206, 207, 215
  logging, 123
  molarity, 67
  permeability, 215, 216
  pollutant, 184
  pollution, 101
  relations, 133
  resources, 118
  retaining capacities, 45
  retention, 14, 74, 78, 81, 119, 158, 161, 184, 210, 218
    capacity, 81, 158, 184
  solubility, 73
  soluble
    carbon nanoparticles, 161
    electrolyte, 221
  solution, 122
  sources, 60
  splitting, 16
  stable macro-aggregate, 214
  storage, 60, 163
  stress, 122
  transpiration, 126
  uptake, 126, 163
  utilization, 211
  vapor, 5
Wavelengths, 130
Weak interactions, 70
Weathering, 122, 215
Weed
  control, 184
  growth, 184
  management, 184
Weeding techniques, 61
Weight, 7, 38, 45, 132, 158, 215, 216
Well-known antioxidant, 76
Wet
  biomass, 157
  chemical technique, 66
  milling, 7, 8, 64
  soils, 193
  torrefaction, 157

Wheat, 102, 120, 126, 129, 132, 136, 161, 181, 184, 189
char, 184
crop, 120, 189
plant, 126, 136
straw, 102, 120, 164, 181, 186
Wide
applications, 47, 69, 81
range, 4, 6, 7, 33, 37, 38, 61, 100, 106, 140, 153, 158, 180, 207
variety, 42, 45, 106
Widespread, 106, 188
benefits, 106
soil pollution, 187
Wilts, 189
Wind, 60, 122
Witnessed, 18
Wood, 1, 6, 7, 34, 38, 120, 136, 163, 179, 180, 182, 189
biochar, 189
waste, 185
Woody
BC, 120
biochar, 38
materials, 192
Workable plan, 191
World
bank, 178
croplands, 123
irrigated land, 119
Worldwide, 119, 223, 226
scale, 185
Wound regions, 141

X

X-ray
diffraction, 10, 120
photoelectron spectroscopy, 10
Xylosidase, 220

Y

Yeast, 61, 68
Yield, 4, 5, 7, 9, 17, 18, 36, 37, 45, 66, 96, 97, 105, 118, 119, 123, 129–131, 133, 135, 136, 153, 154, 156, 157, 160, 163, 165, 180, 187, 190, 191, 194, 223, 225
Young's modulus, 14, 71

Z

Zeolite, 101, 228
Zero, 42, 154, 183
Zerovalent iron (ZVI), 42, 183, 227
Zeta potential, 16, 18, 36, 47, 78, 97, 118
Zinc (Zn), 42, 69, 73–75, 80, 118, 132, 163, 164, 186, 189, 213, 219, 220
nanoparticles, 74, 75
oxide (ZnO), 44, 69, 74, 101, 132, 164, 186, 208, 219, 220
nanoparticles (ZnO NPs), 44, 74, 219, 220
sulfate, 42
sulfide (ZnS), 37, 42, 101, 164, 182
nanocrystals, 37, 42, 182
Zirconium oxide, 67